ROBERT INGPEN

IMPRINTS OF GENERATIONS

THE PROGRESS OF CONSERVATION IN AUSTRALIA

Lothian
BOOKS

Lothian Books
An imprint of Hachette Livre Australia
132 Albert Road, South Melbourne, Victoria 3205
www.lothian.com.au

First published 2006

National Library of Australia
Cataloguing-in-Publication data:

Ingpen, Robert, 1936- .
Imprints of generations.

ISBN 0 7344 0957 5.

1. Ingpen, Robert, 1936- . - Illustrations. 2. Habitat conservation - Australia - Pictorial works. 3. Cultural property - Protection - Australia - Pictorial works. 4. Historic preservation - Australia - Pictorial works. I. Title.

363.690994

Edited by Helen Chamberlin
Designed by Robert Ingpen and Tina Denham
Colour reproduction by Publishing Prepress, Port Melbourne
Printed in China by SNP Leefung

CONTENTS

INTRODUCTION

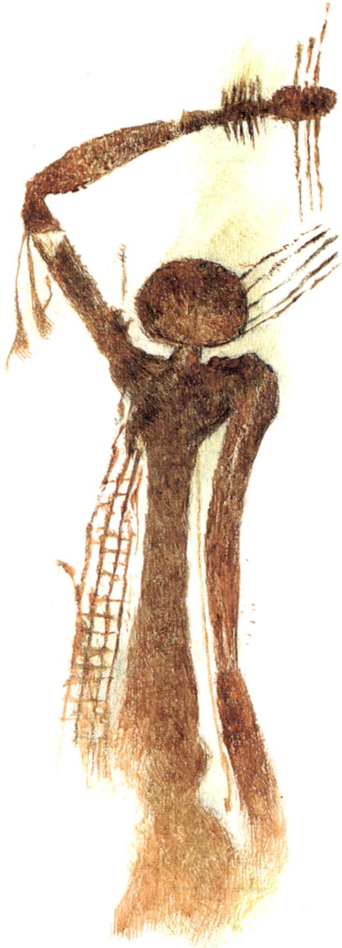

A 'Bradshaw' figure from long ago, drawn on a rock wall in the Kimberley, Western Australia, by an unknown visitor or settler

People have arrived and settled in Australia for a long, long time. We do not know who first came to this great south land, nor how they got here and where they came from. Did the Australian Aborigines choose to come here or were they pushed? Did they leave some place in Asia or further away because, like the modern-day 'boat people', there was no longer a place for them in the land where they were born? Were they sent here, like the first European convicts, because society had no need for them? Did the waves of settlers of this big island really choose to be here, or have they had to fit into it and make the best of what might be described as their 'unchosen land'?

History, as we are taught it, is not what actually happened but what gets written down. Even today, most of what happens escapes the telling because it is too commonplace to be considered worth recording. In his 1998 ABC Boyer Lectures, 'The making of Australian consciousness', novelist and poet David Malouf observed '...the history that is in objects may need to be excavated and made visible before we can experience the richness it represents'.

Imprints of Generations attempts, through pictures and words, to 'excavate and make visible' the richness of some of the places, people and ideas that have shaped our life and culture here in Australia.

As Malouf explains, 'The real work of culture is the business of making accessible the richness of the world we are in, of bringing density to ordinary day-to-day living in a particular place. It is a matter of enriching our consciousness, in both senses of that word — of increasing our awareness of what exists around us, making it register on our senses in the most vivid way; but also of taking all that into our consciousness and giving it a second life there, so that we possess the world we inhabit imaginatively as well as in fact.'

In recent times, over the last 200 years, we have 'flooded' into the inland of Australia only to ebb back again to a more manageable coastline and other favoured pockets. We have built monuments and told stories to celebrate our battles with our harsh environment, and found ways to remind future generations of our impacts and imprints — our failures and fortunes. With the land as a kind of sleeping partner we have created ingenious tools and machines to extend our potential, and sometimes that of nature, so that we might work the alien soils and harvest introduced crops and livestock. We have wonderful dreams about how our country and our future might be, so that when our athletes, scientists, educators and artists briefly capture a moment on the world stage, we applaud ourselves — not so much for what we are but for what we wish to be.

As a nation we continue to race, often carelessly, from dreams to reality, from enthusiasm to despair, from too much to too little, but through all this we are learning. We are sensing that, by daring to become diverse and multicultural, our society runs more risks than others of breaking up and fragmenting. Such threats are not new. As we graduate from a colonial history, with its thoughtless plundering of land, to something more permanent by developing effective new practices to prolong the abundance of our habitat, we embrace conservation.

This book is about conservation — the art of understanding the richness of the world we live in, how humans and nature share, and how we can maintain a constant watch over what we know we have, to ensure that the best of it is not perverted or destroyed but saved for future generations.

The conservation mural reproduced here was painted in 1970 for the entrance to the Rylah Institute for Environmental Research, Melbourne. Each panel depicts a section of natural river bank and shows how nature is altered by human use.

At first there is a true natural wilderness, nature unaffected by people, even though they may be among the creatures who live in it.

The second panel shows that when humans, inquisitive and acquisitive, begin to measure and explore the natural habitat, they are no longer content to simply live within it. They try to understand how it works which, in itself, is another kind of wilderness of complexity.

In the third panel there are signs of human exploitation of the land for profit, without first understanding the effect of their actions. The river bank is eroded and there is weed growth and water pollution resulting from careless agriculture and land use.

Eventually, a final wilderness is reached. The damage to nature is now very obvious, but people often ignore the crisis. They hope that somehow it will adjust itself. But nature moves at her own pace, which is too slow for the impatient lives of humankind.

In the wider world beyond this symbolic river bank, this fourth wilderness is marked by global warming of our environment, species extinction and other consequences of our careless actions.

IMPRESSIONS OF WILDERNESS

The illustration over the previous pages is reproduced from a mural painting for the Arthur Rylah Institute for Environmental Research, Melbourne, which I designed and painted in 1970. Each panel depicts a section of a natural river bank somewhere in Australia. The first panel on the left shows the bank and water in its natural state. The second reveals the mark of humans' attempts to study and record Nature in their first steps to exploit natural resources. The third panel shows how human efforts to force nature to submit to our needs often exhaust and lessen chances of environmental renewal. Then eventually, without intelligent control, this situation deteriorates into chronic pollution and imbalance.

Each of these stages can be applied at any habitat or section of our natural landscape. They can also be viewed in another way — they can be used to demonstrate different applications of the word 'wilderness'.

'Wilderness' is mostly used to describe an uncultivated and still wild region, but it can also mean a confused assemblage of things. Used metaphorically it can mean to be 'out of office', as to be in a political wilderness where you are no longer truly effective.

Revolutionary changes in thinking about wilderness and our natural surroundings began about 150 years ago, during the great upheavals of the Industrial Revolution. That revolution, which began almost a century earlier, altered the social structure of the western world and created a smoky wilderness which eventually spread over much of Europe and North America. Previously, wealth had been inseparable from the produce of the land and the richest men were the great landowners. They became even richer when their land yielded coal and iron as well as food and fibre, but the huge industrial boom also brought wealth to a new breed of technologists and entrepreneurs. And there were greater opportunities for everyone, including those who flocked to find fortune in the new gold diggings in Australia.

Charles Darwin, naturalist

In the midst of this upheaval, two men published totally different and yet somehow related theories, both associated with wilderness. They were Charles Darwin in England and Henry David Thoreau in America.

Darwin's great book *The Origin of Species* first proposed, to a startled and often resentful audience, that human beings were only a part of the whole natural world and not godlike creatures who had a right to use that world solely for their own benefit. Only as a part of a long complex chain of evolution had they earned their right to survival.

Henry Thoreau, poet

Thoreau's book *Walden*, is seen by some as the first cry of the conservationist against humans' deliberate divorce from the natural world, and against the perversion of human nature which came from regarding the land as nothing but a source of profit. Thoreau, an eccentric poet, was a surveyor by trade, with no training in the sciences. *Walden* was his published account of life alone for twenty-six months in the natural wilderness of the woods surrounding Walden Pond in Massachusetts. He observed that this wilderness area was really controlled by natural laws largely unknown to mankind, and wrote, 'If we knew all the laws of Nature, we should need only one fact, or the description of one actual phenomenon, to infer all the particular results at that point. Now we know only a few laws, and our result is vitiated, not by any confusion or irregularity in Nature, but by our ignorance of the essential elements in the calculation…'

The writings and observations of both Darwin and Thoreau have taken a long time to affect the thinking of the average person. Only in the last fifty years have the subtle effects of Thoreau's thinking infiltrated our feeling about our surroundings in the natural world, and given us hope for the future.

One of the first men to see the potential of Australia as a field for ecological study and conservation management was the American George Perkins Marsh. In 1864, when the impact of the gold rushes already had caused enormous environmental damage, he wrote in his book *Man and Nature: Physical Geography as Modified by Human Action* that: 'Australia is, perhaps, the country from which we have right to expect the fullest elucidation of problems [concerning ecology]... the peculiarities of its fauna, its flora, and its geology are such as to have excited for it the liveliest interest of the votaries of natural science; its mines have given its people the necessary wealth for procuring the means of instrumental observation, and the leisure required for the pursuit of scientific research; and large tracts of virgin forest and natural meadow are rapidly passing under the control of civilized man. Here, then, exist greater facilities and stronger motives for the careful study of the topics in question than have ever been found combined in any other theatre of European colonisation.'

Matthew Flinders, navigator

The 'careful study' he advocated had already begun by then in Australia and continues now. The earliest recorders included botanist Sir Joseph Banks and navigator Matthew Flinders, whose great work, *A Voyage to Terra Australis*, was published exactly fifty years before Marsh's book. By that date, 1814, there were already signs that Australia would develop into something more than a dumping ground for the dregs of English prisons, and Flinders'

Even after over two hundred years, Matthew Flinders' maps of the coastline of Australia are among the most accurate ever made. They can be relied upon by navigators of the most hazardous waters such as those of the Great Barrier Reef off Queensland. Here is a detail of the coastline near the Whitsundays, as mapped in 1802 on H.M. Sloop Investigator.

ISLES
THE NORTHUMBERLAND ISLES
CAPTAIN COOK. 1770.
C. Hilsborough
Point Slade
C. Palmerston
PERCY ISLES
Thirsty Sound
BROAD SOUND
SHOAL-WATER BAY
North-Point Is
Flat Is
C. Townshend
Strong-tide Passage
Harvey's Is
Port Bowen
C. Clinton
C. Manifold
Double Mt
Pine Mount
Hummock Mt
Deptford struck
June 2.1770.
June 1.1770
May.28.1770.
Aug. 25.1802.
Oct.5.1802.
High breakers
Very high breakers
The tops of this extensive ridge of high hills appeared craggy in many places, & generally barren; but the Northern sides are woody.
tude 149°E. of Greenwich.
30'
150°.
30'
151°.

publication of his own meticulous maps and records of this new land aroused great interest among the politicians, scientists and navigators of that era.

Much of what he wrote is still relevant today, 200 years on. For example his advice to sailors approaching the Great Barrier Reef: 'The commander who proposes to make the experiment must not, however, be one who throws his ship's head around in a hurry so soon as breakers are announced from aloft; if he does not feel his nerves are strong enough to thread the needle, as it is called amongst the reefs, whilst he directs steerage from the masthead…'

John Gould, naturalist

Shiploads of immigrants arrived in the new land which Flinders had defined on his charts of the coastline, and they stepped ashore wide-eyed and wondering. Many of the earlier artists were baffled by the subtle differences between Australia and their homeland. The trees appeared more monochromatic, the skies more intensely blue, the sun more startlingly brilliant, and the shadows cast far blacker than they had ever seen in Europe. Instead of familiar flora and fauna there were strangely beautiful flowers, exotic, gaudy and noisy birds, animals which seemed totally unbelievable, and a race of humans like none other in the known world. The first Europeans hardly believed what they saw and painted the eucalypts to resemble English oaks, the Aborigines like classic Romans, and the kangaroo like a cross between a deer and a dog.

But when more skilled observers such as John Gould and Ferdinand von Mueller arrived in Australia around the 1850s, they were enchanted, not dismayed, by the differences. They set to work to record them accurately. Gould published his *Birds of Australia* in seven volumes, between 1841 and 1848, and its more than 600 lithographic illustrations have never been surpassed for accuracy and beauty. Between 1845 and 1863 he published *The Mammals*

Echidnas from Gould's
The Mammals of Australia

of Australia and both series offer a priceless record of Australian wildlife from the days when the various birds and animals lived in greater numbers than they do today.

Ferdinand von Mueller, botanist

Von Mueller performed a similar service for Australian botany. From 1847 onwards he collected specimens throughout South Australia, Victoria and Northern Australia, often penetrating areas where no white person had travelled. In 1857 he was appointed Director of the Melbourne Botanic Gardens, and by 1873 he had established one of the finest botanic collections in the southern hemisphere. Eventually he had collected and noted an estimated 350 000 botanic specimens.

Samuel Thomas Gill concentrated upon recording a different type of species: the white humans who were establishing themselves in Australia between 1840 and 1880. His drawings and watercolours have provided us with an accurate record of life in early Adelaide on the copper mines of Kapunda and Burra in the 1840s and the goldfields of Victoria in the 1850s. His talents and endurance have endowed us with a graphic pictorial record of social life, costumes and lifestyles during those formative years of our nation.

Samuel Gill, artist

Gill died in 1880, just before a new generation of Australian writers emerged to record the facts of life in colonial Australia. Their doyen was Henry Lawson, and they turned away from the romanticism of the convict era, bushranging and goldmining to describe what life was really like for the battlers of the bush and the city. Lawson, like Joseph Furphy and others of his period, knew such a life from personal experience and he founded a tradition of realistic writing which continues to this day. His insight into the starkness of life as it was lived by the people of the outback is displayed in such verses as those supposedly spoken by a farmer's wife, beaten into insensitivity by endless labour and disappointment:

Deep Sinking, Ballaarat 1855, *S.T. Gill, Lithograph*

By the end of 1852 S.T. Gill had established himself as the 'artist of the goldfields' — a title which tended to obscure the value of his earlier artistic work in Adelaide and the copper mines of South Australia. Gill died a derelict in 1880, and was in his last years selling and raffling his drawings in hotels for the price of a drink.

The crops have withered from the ground,
The tank's clay bed is glarin',
But from my heart no tear or sound,
For I have got past carin'.

Henry Lawson, writer

Early in the twentieth century another Australian wrote the truth about an event that is still subject to an immense amount of romanticism: the First World War. C.E.W. Bean was a well-known author and journalist who worked for the *Sydney Morning Herald*. Fellow journalists nominated him for the post of Official War Correspondent, and in this capacity he visited every battlefield on which Australians were fighting and saw the picture first-hand. He talked to men of all ranks and branches of the forces, and to politicians at home, in his endeavour to write 'the plain and absolute truth'. After the war he was the editor and main author of *The Official History of Australia in the War of 1914–1918.*

Bean's great work was one of the most important steps forward in the process of Australians recording their own history, which is vital if we are to see our true place in the scheme of things.

In more recent times the great historian Charles Manning Hope Clark published the volumes of his remarkable *History of Australia*. His discipline as an historian and his moralistic passion are blended in his refreshing and challenging interpretation of historical sequence and fact, so that they form a panorama of our country. In his Boyer Lectures of 1976 he quoted the Henry James epigram 'a man gives what he has: the rest belongs to the madness of art' and commented, 'Perhaps in our country so far we have been too timid to surrender to that madness.'

C.E.W. Bean, historian, author and journalist

Christmas Eve, the Western Front, World War I

'On Christmas season during World War I, in the most dreadful days of all, between a thousand and a thousand deaths, men walked towards each other. Enemies, wondering, sharing "the air of an old Austrian song — Silent Night, *and coming closer, saw not enemies but themselves, in that redeeming dawn.'*

Patricia Crampton

FOLIO ONE IMPRINTS OF GENERATIONS

Limestone scallop fossil

Nearly everything leaves an imprint to mark its having been once, and in some cases, to make sure it is not forgotten. There are imprints, marks and tracks everywhere around us. They have been made by people, animals, birds, insects, plants, even the sun, wind and waves. The prints of a kangaroo, a crow, or a crab left in the sand can tell a skilled tracker much more than just what creature walked there. They can guess by the tracks as to where, and where from, or even why the creature was moving. Crazy marks and carved tracks on leaves and timber signal the feeding of grubs and caterpillars, as each makes its own distinctive imprint as it seeks and chooses its food and lifestyle, then moves on.

Imprints are personal and sometimes private. The annual growth rings within the trunk of a tree tell its age, and keep the secret until the tree is cut down or decays.

People have fingerprints that make a mark that distinguishes them from all other people. Fish have scales with growth markings that tell of their progress through life. Deep within the earth's rocks lie fossils of plants and creatures that have lived and, seemingly, want to be remembered millions of years after they have lived.

After so many years, Australians as individuals or in groups are slowly accumulating a rich treasury of wisdom which is helping to guide our routines, form our habits and provide acceptable explanations for commonplace events and the complexity of nature's moods. Our scientists are observant and rational, and continually monitor our progress to help us judge when it is time to adjust our pace or alter our direction. Our educators and artists remind us that humans are so infinitely adaptable that we can actually live happily with change. Our environmentalists, the individuals who bridge the sciences and the arts, plead with society to ensure that change will be for our benefit and not for our destruction. That no change at all will occur is unthinkable. Somebody whimsically observed that wilderness is a place where if you stand still something will eat you!

We must move carefully through our wilderness keeping a constant watch on what we change, what we keep and what is thrown away, and why. Our heritage is the sum of how well we can do this and account for ourselves for the benefit of those yet to be born.

But people leave the imprints that are most varied and unexpected impressions of their being on earth. Each passing generation leaves its imprint on each new generation. We of today inherit a way of living which bears the consequences of the actions of our forefathers, then we pass on to our children a way of life further changed. This pattern is our heritage that identifies and details our progress as each generation leaves its imprint to remind us of them and their age.

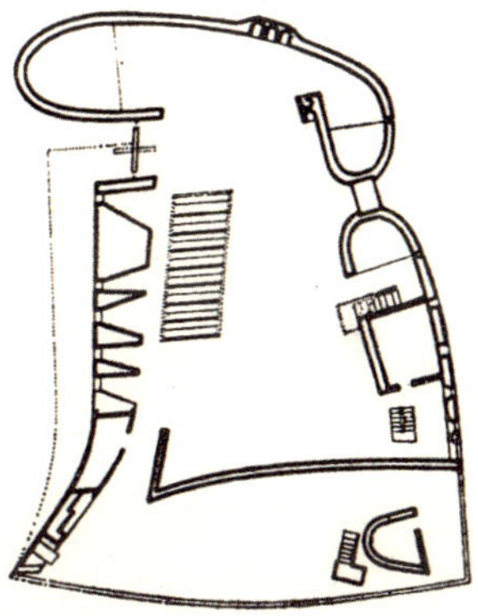

'Footprint' of the famous Notre-Dame-Du-Haut, Ronchamp Chapel in France, by architect Le Corbusier.

This heritage is our buildings, tools and implements, our ideas and dreams, and our contact with the natural surroundings. The imprints they make tell of who we are, and who we would like to be as we live and grow and change.

The floor plan of each building is as distinctive as a footprint. It is the essence of each building, reflecting the design and building materials, its purpose and performance, and so long as it remains, its mark for us to track. This is the same for all buildings that have once been made, that are constructed today, and those that will be built in the future. It is the same for castles and churches, temples and townhouses, for banks and barns, and simple dwellings and shelters. Each has its distinctive imprint on the ground and on us.

The marks made by a stone-age axe that trace an ageless totemic story in stone or wood stand at the beginning of a chain of impressions that remind us of our origins by telling others about the source of our ideas and beliefs. Likewise, the imprint of the creative arts and crafts made by somebody, often unknown, leaves its impression on those who see, touch, hear or wear the creative artefact.

Then there is the imprint of a story or a song first told long ago around a hunter's campfire, then later written down on paper and illustrated with a drawing. These leave revealing and complex accounts of the human imagination, invention, scientific thought and change, to be interpreted and applied to be helpful as our world changes with each age.

'...Our notions of law and harmony are confined
to the instances which we detect;
but harmony which results from a far greater
number of seemingly conflicting, but really
concurring laws which we have not detected
is still more beautiful.'

Henry Thoreau, *Walden*

A coral rock community, upper and underside, drawn on Heron Island in 1980 to celebrate World Heritage nomination for the Great Barrier Reef.

The Murray River Passengers' Guide Book

Issued by

Murray Shipping Limited

21, Steamship Buildings,
Currie Street, Adelaide

An extract from The Murray River Passengers' Guide Book *issued to riverboat passengers in the early 1900s. The following report is by the Scottish Agricultural Commission which, in 1910–11, travelled down the River Murray from Mildura in the paddlesteamer* Marion *in December. They wrote:*

'As we are borne down the stream we are conscious of an infinite variety of detail lightening up a scheme of Nature generally sombre and a little monotonous. Sometimes the eucalypt trees show up a rose-red, sometimes a mottled sea-green, sometimes a pale blue toning off into a slaty-white. The banks at one part are densely lined with brightly foliaged willows; at another forest thins out, and we have the flash of a shining backwater, at another we pass through cliffs of sedimentary rock, composed of friable sandstone and clay. Nor are the woods without life; a herd of emu stalk unhurriedly across the field of vision; a bird breaks forth in song that might do honour to a leafy English lane; cockatoos and other winged creatures of bright plumage arise from thickets, re-alight and chatter. By night the river suffers a mysterious change, the shadows and reflections of the trees being cast in such a way as to compose a dim underworld which might to the fancy be inhabited by gnomes and nymphs. We are not surprised at the fascination the Murray exercises, either as potential wealth or as river scenery preserving all features of Nature's own handiwork, unaltered as yet by man and inviting the recreational spirit of man.'

The illustration is of an old river gum marked long ago by a number '33' to indicate to riverboat captains how far they were from their next port on the River Murray.

'Storm-Boy lived between the Coorong and the sea.
His home was the long, long snout of sandhill and scrub
That curves away south-eastwards from the Murray mouth ...
... Storm-Boy lived with Hide-Away Tom, his father.
Their home was a rough little humpy made of wood
and brush and flattened sheets of iron ...
... It was hot in summer and cold in winter, and it
shivered when the great storms bent the sedges and
shrieked through the bushes outside.
But Storm-Boy was happy there.'

Colin Thiele, *Storm Boy*

MODELS AND REALITY

Change is inevitable, and we know that we must ask ourselves regularly 'What kind of degree of change, from one time to another, is appropriate and acceptable?' As progress becomes more rapid, changes occur almost before we know they are there, and they affect our lives in ways we often cannot predict.

With science to help us we can plan to understand change and the consequences by developing models. In biology, for example, a model may be a theoretical representation of a function of a biological system. It can also be a physical construction showing the organs, tissues and cells clearly separated from the surroundings which would normally obscure them. Or a model may be developed by mathematicians, using symbols to represent elements, events and relationships, to assist calculations by diagrams on paper or within computers.

A model may even consist of words alone. A good poem, which mirrors our emotions and aesthetic responses and concepts of our changing society, is in some ways a model which helps us to understand ourselves. It can help to expand our view of ourselves in society and our natural environment.

In his book *The Landscape of Australian Poetry*, Brian Elliott created a word model which makes us think about how we can know Australia. 'The first need in a new country or colony must obviously be in one way or another to comprehend the physical environment. At first the urge is merely topographic, to answer the question what does the place look like? The next is more detailed and ecological: how does life arrange itself? What plants, what animals, what activity, how does man fit in? The next may be moral: how does the place influence people? And how, in their turn, do the people make their marks upon the place? Next come the subtler enquiries: what spiritual and emotional qualities do such people develop in such an environment? In what way do the forces of nature impinge upon the imagination? How do aesthetic evaluations grow? How may poetry come to life in such a place as Australia?'

These questions attempt to simplify the greatly complex real situation of humans and nature in Australia. They aim to assist us to find our present place in that situation, so that we might search with more confidence for answers to the greater question: that concerning the amount of change that is appropriate and acceptable.

Banksia serrata *after a drawing by Sydney Parkinson in 1770, while with the James Cook expedition at Botany Bay.*

MODEL '...What does the place look like?'

Our pioneers looked at Australia, but they did not really see it. For much of the time they tried desperately hard to force the country to make their dreams come true: to convert it into something it could never be, and what was called 'A Great Britain beyond the seas'.

They cut down the gum trees and planted oaks, ash and pine. They built flat-faced Georgian-style houses which blistered in the sun, and ate roast turkey and boiled puddings for Christmas dinner on days when the sun almost boiled water in the tank.

The land confused, unsettled and frightened them by its seeming irrationalities. In Victoria, a pioneer settler wrote of the Mallee country. 'Then it was discovered that tracts which had been passed over as barren in the summer had a peculiar value in the winter, and in fact it gradually became apparent that they were second to no district in their capacity for producing fat stock, the fattening seasons, however, being different.'

Fattening sheep in winter instead of summer seemed weird. At other times they found that 'a fertile region which had presented itself to a delightful view' had changed into 'an arid waste, destitute of either grass or water'.

The colonial settlers devoted an immense amount of physical, mental and spiritual energy to trying to turn Australia into their own picture of what the land should be, and hardly saw what the place actually looked like and why it was so. Slowly, they discovered that they had to adapt themselves to their surroundings, and the only way they could prosper was by evolving new and different methods of agriculture.

But many learned this lesson too late. Their memorials are abandoned homesteads, perhaps nothing more now than a tumbledown chimney remaining, and the introduced pines which they once planted as a windbreak or shelter, or as reminders of home far away.

Francis Ratcliffe, scientist

In 1947 Francis Ratcliffe, a biologist who grew up and studied at Oxford in England, wrote *Flying Fox and Drifting Sand*. In this account of his work as an applied ecologist for the CSIRO, he reviewed his travels in the northern tropics of Queensland, and in the outback of inland Australia, or 'the kingdom of the dust' as he called it. In the final paragraph of his book, which marked him as the founder of conservation thinking in Australia, he observed 'The conquest of the Australian inland demanded fortitude and endurance, self-reliance and faith; nor did the need for these qualities disappear when the conquest had been achieved. At times during the last few years the need has probably been greater than ever before, for hope must have given place to disillusionment in many hearts. Australians have every reason to be intensely proud of their record in settling the great spaces of the inland. They are to be blamed only in that they seem to have done the job too thoroughly.'

That was written exactly one hundred years after George Perkins Marsh had made his observations about Australia and its potential as a place for the study of ecology.

I praise the scoring drought,
the flying dust,
and drying creek,
the furious animal,
that they oppose us still;
that we are ruined by
the thing we kill.

Judith Wright,
Australia 1970

Even after two centuries, we do not really understand or comprehend the physical environment. In an age of spectacular technological achievement when humans can reach into space, and satellites can spy the formation of cyclones, we have little genuine knowledge or response to Australia weather. In the driest, harshest and most exposed of all continents we are struggling to evolve a system to accurately predict the weather for more than two or three days ahead.

At various times during the last thirty years or so, scientists have spoken of the need for a true 'environmental frame': an organised grid of recording stations covering all our lands and the continental seas, as well as from satellites, that harvest weather information continually. To this scientifically gleaned data should be added the folk knowledge accumulated and incorporated in storytelling by such people as the Aborigines, who have lived more closely with the environment than anyone else, and the farmers and rivermen and fishermen who have learned by painful experience to 'read the signs of nature'.

Meteorology, the science of weather, may have to be expanded to accept ways currently thought to be unscientific and make-believe. Already we draw

The sea-god 'El Niño', adapted from an image on ancient Inca pottery in Peru, South America

upon folklore and myth when we apply the title 'El Niño' to a meteorological oceanographic episode that impacts on our weather patterns. The term 'El Niño' (the child) comes from a Peruvian Inca story of about five hundred years ago when there was no science at all, and phenomena were explained by storytelling. Every so often the ocean current off the coast of Peru would become too warm for the fish it contained, and they would have to migrate. The wildlife and people who needed to catch the fish, particularly huge stocks of anchovy, went hungry. The Inca explained this episode by saying that the sea god, El Niño, who roamed the Pacific Ocean close to Peru in his serpent boat, and with his pelican slaves tethered to him to catch his fish, became greedy. He tried without success to catch much bigger fish, and became so angry at his failure that his temper raised the temperature of the ocean around him. After a while, when he calmed down, the water returned to normal, and the schools of small fish returned.

MODEL 'The next need… is detailed and ecological. How does life arrange itself? What plants, what animals, what activity, how do humans fit in?'

A stunning object lesson of how humans tried to 'fit in' to the ecology is presented by what happened at Tower Hill in southern Victoria, firstly in the 1850s and then again one hundred years later. Australian painter Jean Eugene von Guérard painted the Tower Hill crater and island some time in 1855. He painted it as it appeared in those days with such accuracy that native plants in this 'treasure house of nature' could easily be identified. By 1870, the scene had already changed radically for the worse. Burning, grazing and cutting firewood had depleted much of Tower Hill's natural vegetation,

Tower Hill, 1855, *Eugene von Guérard*

Tower Hill, an extinct volcano in the western district of Victoria, looked like this at the time of pastoral settlement. By 1870 the scene had changed radically. Burning, grazing and timber cutting had destroyed much of the vegation. With illegal 'squatting' on the island and quarrying on the crater walls, the wildlife was almost eliminated and the beautiful lake almost dry.

With the re-creation begun in 1960, we may eventually 'see' Tower Hill as it was when this painting was done so long ago.

the walls of the extinct volcano were quarried, and the island was used as a common for grazing. Illegal squatters, vegetable growers and graziers eliminated what wildlife remained around the edge of the beautiful lake, which almost dried up.

Succeeding generations tried to repair the damage by introducing European plants, animals and birds, but such 'experiments' were equally disastrous.

Alfred Dunbavin Butcher, scientist and administrator

In 1960, the Victorian Fisheries and Wildlife Department, under the guidance of Alfred Dunbavin Butcher, took up the challenge of trying to restore Tower Hill to its natural state. Butcher and his scientists knew it would take many years of careful effort to reverse the damage humans had done to the island, lake and walls of the great collapsed volcano. The success of the restoration plan lay in gathering as much information as possible about what had been done in the past, as well as what might be achieved now and by future generations.

After forty years of careful restoration the outlook is promising. Massive plantings of native trees, shrubs, grasses, rushes and herbs of the types shown in von Guérard's 'guide' painting have replaced the originals stripped away long ago. In the beautiful tree fern gullies, the replanting must be accompanied by the restoration of the special habitat communities of insects, birds and animals.

We are now beginning to see Tower Hill as von Guérard saw it so long ago. The re-creation is being carried out at great cost and so the public can visit this experiment in how we can learn to at last 'fit in', to understand how native life arranges itself, even after humans have made such a mess.

On a much bigger scale the vast Murray-Darling-Murrumbidgee river system is far too complex for Tower Hill-type restoration.

For a long time individuals and associations concerned with river conservation have been sounding a warning note. So far it has had little effect upon the network of government authorities and vested interests who dictate the use and misuse of the inland river system.

Humans have not only affected the flood plains by clearing, they have introduced plants and animals which are changing the rivers. European carp are muddying the waters and reducing the plant life, thus reducing the diversity of native life. Introduced grazing animals, both feral and farm, cause erosion and muddy the streams and billabongs. Agricultural practices have greatly increased the quantity of salt flushed into the rivers, and in the Riverina the introduced water hyacinth is clogging waterways and reducing the oxygen content of the water, killing many native animals and plants. In some places the river water has become so salty, because of irrigation and minerals being washed out of the soil as water seeps back into the rivers, and so polluted with artificial fertilisers, that it can no longer be used even for irrigation.

The Murray-Darling-Murrumbidgee river system represents one of Australia's greatest natural resources, of enormous environmental importance, to say nothing of its integral links with our historical and cultural heritage.

Perhaps the problem is too great for us to overcome just yet. Maybe, if it is not too late by then, we will have to wait for the next generation to seriously apply themselves and their imaginations to the solution. Or maybe the next part of the model should be brought into play.

Model **'The next may be moral: how does such a place influence people? And how, in their turn, do people make a mark upon the place?**

The Ya-itamthang Aborigines once journeyed each summer from the Murray river plains up to the Bogong High Plains, to feast upon the plump brown bodies of bogong moths. Cattlemen drove the Aborigines away and took stock up to graze on the rich grasses above the alpine snowline. Prospectors crossed and re-crossed this high wilderness, timbermen plundered the age-old stands of mountain gum, and engineers built dams and blasted tunnels to force the mountain cascades into hydroelectric schemes.

All these invaders left their imprints on the country, but its influence upon them may have been only that of promising some kind of gain. Among the marks of white people are the alpine drovers' huts and refuges which still stand at intervals along age-old trails. They now wait for another night or snow storm when some traveller will need shelter from the world around. These buildings, like many others of our more isolated wilderness areas, may appear very basic and abandoned. Yet they mostly suit their natural surroundings better than some modern constructions.

Now, large areas of the Bogong High Plains are reserved as a national park. Bushwalkers and skiers traverse the high country, and its influence upon them is certainly the invigorating peace and refreshment of wide vistas and open skies. Hopefully, all they will take from the place will be photographs and lasting memories, which leave no lasting mess.

Far from the mountains, the *Polly Woodside* lies like a living symbol of another kind of restoration of our environment. In her case, she signifies the

'Twas they who rode the trackless bush in heat and storm and drought;
'Twas they who heard the master-word that called them farther out;
'Twas they that followed up the trail the mountain cattle made
And pressed across the mighty range where now their bones are laid.

A.B. 'Banjo' Paterson, *Pioneers*

historical environment in which our forebears lived and worked. The *Polly Woodside* lies in the once busy 'Pool of Melbourne' which was the furthest point that sailing ships could venture up the Yarra River at the end of their voyages from all the oceans of the world. From its deck you can see the business towers of modern Melbourne and the new Docklands developments. Twenty-five years ago this barque was the only sailing ship afloat in Australian waters. Built in 1885, and reduced to no more than a hulk, she was then restored by the National Trust of Victoria. Today there are a number of sailing 'tall ships' afloat, and more, like the re-creation of James Cook's *Endeavour* to continue to remind us of our origins.

As a visitor to the *Polly Woodside* you can smell the tar, varnish and rope. You can hear the creak of timbers and the breeze in the rigging. You stand on a ship in which every detail was hand-crafted from timber, iron, cordage and canvas, in ways which our computer age would regard as impossibly uneconomical.

The *Polly Woodside* is a 'model' now which may be regarded in several dimensions. She serves to remind us of the influences of the sailing ship on Australian development; she allows us to see the detailed craftsmanship of an earlier age. She shows us too that a whole society existed and maintained itself for centuries before petroleum first gushed from the disputed deserts of the Middle East… and that people will still be able to continue, perhaps a little changed, even after the oil wells have been emptied.

In another dimension, the *Polly Woodside* demonstrates the enthusiasm of spare-time preservationists who have done so much to save and maintain memorials to our past. Such people, all of them amateurs in the sense that they do work which they love, stand up to the great juggernaut of progress

which would destroy anything regarded as 'unprofitable' because it does not make money.

Australians are 'making their mark upon the place' and the difficulty is to establish a logic as to which of these 'marks' should be preserved, and which removed to make way for other and possibly better things. Preservation simply for its own sake may not always be logical, and there is always the danger that human bloody-mindedness may cause stubborn determination either to preserve or demolish. The difficulty, too, is that the object which we now see as ugly and useless, because it is an outmoded modern construction, might be regarded as a priceless historical relic by people of a century in the future. The *Polly Woodside* was only one of thousands of ships in her period: ocean-going semi-trailers driven by the wind and built solely for profit, then broken up when steamships made them redundant. It is a miracle that even her derelict hull escaped the shipbreakers.

It is worth remembering that the tower from which the first United States astronaut was launched into space has long been demolished on the grounds that it was too expensive to maintain as a historical monument.

MODEL **'Next come the more subtle enquiries: What spiritual and emotional qualities does such a people develop in such an environment? In what way do the forces of nature impinge upon the imagination?**

Six treasures hang on my studio wall. Six rusty artefacts, which many of my visitors feel are the kind of junk one tosses into the non-burnable rubbish. They have no possible practical use in the modern world, but I cherish them

because each marks a time, a significant place, a reserved memory that I keep and value.

One is part of a stirrup that I found in the red dust around the historic One Tree Hotel, sometimes called 'Hell'. The old building stands alone on the great saltbush plain between Hay and Booligal, and marks the intersection of two overland stock routes across New South Wales and beyond.

Alongside the stirrup hangs a pair of crude stonemason's callipers, found in the crumbling mortar of the convict-built foundations to Australia's first woollen mill at Lithgow, NSW. And a little iron hook, given to me by Mrs Lincoln of Lincoln's Place on the Bellarine Peninsula of Victoria. It was once on a chain that hung from the mulberry tree outside her home, which a century or more ago was a cheese factory. Mrs Lincoln hung her faithful pitchfork, which she called 'Johnny Grigg', on that hook while she rested.

There are three chain links, rusty and time-bitten. Once they secured the chapel door in Drunkards Gully near the South Australian pioneer seaport of Robe. The chapel marks the campsite of countless bullock teams bringing wool to port from the inland along romantic Baghdad Road.

The last two treasures come from places far away. There is a piece of shrapnel found in the sands of the steep cliff of Anzac Cove in Turkey, and a segmented silver fish from the ancient seaport of Calloa in Peru. Our son Tom found the fragment on his visit to the Anzac site and brought it home for 'my wall'. I was given the piece of Spanish ornament by a Peruvian biologist who I was working alongside in a Marine Institute in Peru in the early 1970s.

Individually and collectively, these minor treasures speak to me constantly of the true builders of our nation; of dusty mobs of sheep and cattle ambling along stock routes; of early industries and commercial efforts now long abandoned; of the sacrifice at war; of celebration; and of the millions who came to find refuge in this unchosen land from something intolerable that happened in their homeland — the refugees of every time and culture.

We accumulate our treasures either as investment or as physical reminders of an emotional experience. My treasures are of the emotional kind.

In his book *The Australians*, 1966, George Johnston wrote 'If material truth and symbolic poetry were to define a national memorial, then Australia's monument might fittingly be the preservation of just one of the old selector's slab cottages with its roof and chimney of corrugated iron, and ringed water tank, and walls and outhouses textured with a pasted collage of old newspapers and country magazines'. I would go one step further and nominate an actual place, not a hut, but a pub — the One Tree Hotel between Hay and Booligal on the Cobb Highway.

Douglas Mawson, explorer

A close second, by my standards, would be Douglas Mawson's hut at Commonwealth Bay, Antarctica. Each summer snow-thaw partially uncovers the main base hut on Cape Denison of Sir Douglas Mawson's 1911–14 Australasian Antarctic Expedition. On more recent visits, polar scientists have recovered from the hut such small items as reels of cotton, candlesticks holding partially burnt candles, and tins of gramophone needles. They, with the passing of time and the forces of nature, impinge upon our imagination to recall life in the base hut of that ill-fated and heroic expedition. After the tragic death of Mawson's colleagues, Lieutenant Ninnis and Dr Mertz, he walked back alone to the base over 150 kilometres of ice plateau.

FOLIO TWO IMAGES OF OURSELVES

This folio of five old places reflects Manning Clark's statement of thirty years ago. We can fashion dreams around the stones and timber of these abandoned buildings from an earlier time. Those who built them had dreams of their own, and perhaps we can learn from these.

'In the nineteenth century Australians enjoyed the reputation of being in the vanguard of human progress; they were often pioneers and rather boastful about it. By contrast, in the twentieth century Australians seem to have missed the bus carrying humanity into the future. Australians are no longer pathfinders, but rather men of the army of the rear.'

Manning Clark, Boyer Lecture 1977

ONE TREE HOTEL

Cobb Highway, Central New South Wales

A hundred years ago stockmen, shearers, drovers and drifters roistered in this pub they knew as 'Hell' on the stock route between Hay and Booligal. They washed the dust of the track from their throats, swapped yarns and slept it off in 'swags' on the ground outside. Now, as cars and trucks speed along the Cobb Highway not far from this memorable shanty pub, built of timber frame and walls with iron roof and verandah, few people stop to wonder at its story.

Mawson's Expedition Hut
Commonwealth Bay, Antarctica

Each summer snow-thaw partially uncovers the main base hut, on Cape Denison, of Sir Douglas Mawson's 1911–1914 Australasian Antarctic Expedition. This drawing was adapted from photographs taken by Dr Phillip Law in 1962. In recent times visits by polar scientists and restorers have recovered from the hut such items as reels of cotton, candlesticks holding partially burnt candles, and tins of gramophone needles. They recall life in the base hut of that ill-fated expedition.

After the tragic death of Mawson's colleagues, Lieutenant Ninnis and Dr Mertz, he walked back alone to the base hut over 150 kilometres of ice plateau, an heroic feat of human endurance and an epic of polar literature.

MOUNT BISCHOFF TIN ORE PLANT

Waratah, Western Tasmania

On 4 December 1871 James 'Philosopher' Smith discovered the richest deposit of tin the world has ever known. Mount Bischoff was literally a mountain of tin. Within a year or so, at the foot of the mountain in a region saturated by almost perpetual cloud and rain, Heinrich Kaiser supervised construction of this massive ore reduction plant on the side of a ravine. The plant began to process the ore, which was then carted to Burnie by fifty miles of horse rail laid through the mountains. In June 1883, electric light was turned on in the complex for the first time in Australian industrial history. The electricity came from Tasmania's first hydro-electric scheme. The lights stayed on for the next sixty years, until the lode of tin oxide showed signs of exhaustion.

Stud Ram Pens

Wanganella, Riverina, New South Wales

Few people have heard of 'Emperor' and 'Old Grimes': two merino rams of the 1860s. The brothers George and Fred Peppin built these thatched pens to shelter the rams from the hot Riverina sun. Here, by Billabong Creek, they used the two rams to breed the Australian 'Peppin' merino that became a dominant influence on the wool industry for a century. But mystery still surrounds their success. Basil Clapham, the studmaster at Wanganella in later years, believed the English long-woolled Lincoln ewes introduced by the Peppin brothers were secretly mated with the 'pure' merino rams and destroyed, after the lambs were born, to hide the evidence.

Ape Cage

Royal Melbourne Zoo

In 1893 the Zoological Society of Victoria said of the monkeys confined in this cage that they were 'very spiteful and savage in captivity, and when a pair of them are confined in one cage the male continually bents his ill-temper on his unfortunate wife...' The cage is now obsolete but retained as a reminder. In the modern zoo each animal is given something that more closely resembles its natural habitat. Crowded cages and oppressive pits are replaced by pastures and tree-top vision. When fences are needed they are camouflaged by appropriate shrubs and trees.

Only abandoned remains mark the sites and leave the imprint of places that could well have been monuments to our national growth and effort. Places like the tin ore plant at Waratah, Tasmania, where, on 4 December 1871, James 'Philosopher' Smith discovered the richest deposit of tin the world has ever known. Mount Bischoff was literally a mountain of tin. Within a year or so, at the foot of the mountain in a region saturated by almost perpetual cloud and rain, Heinrich Kaiser supervised construction of a massive ore-reduction plant on the side of a ravine.

The plant began to process the ore, which was then carted to Burnie by 50 miles of horse rail laid through the mountains. In June 1883, electric light was turned on in the complex for the first time in Australian industrial history. The electricity came from Tasmania's first hydro-electric scheme. The lights stayed on for the next sixty years, until the lode of tin oxide showed signs of exhaustion. By that time the mine had paid dividends totalling £2.5 million.

George Peppin, pastoralist

Another 'nerve end' in reminding us of our heritage was nearly lost forever, but thanks to far-sighted individuals in Deniliquin, New South Wales, the Stud Ram Pens of Wanganella not far away have been moved to a safer place and faithfully restored for visitors to see. Few people have heard of 'Emperor' and 'Old Grimes': two merino rams of the 1860s. The brothers George and Fred Peppin built the thatched pens to shelter them from the hot Riverina sun.

Here at Wanganella by Billabong Creek, they used the two rams to breed the Australian 'Peppin merino', which became the dominant influence on the wool industry for a century. But mystery surrounds their success. Basil Clapham, the legendary studmaster at Wanganella in later years, believed that English long-woolled Lincoln ewes were secretly mated with the pure merino rams, and destroyed after the lambs were born in order to hide the evidence.

Whatever happened, the Australian wool industry can be grateful to the Peppin brothers, who summed up their unconventional breeding programme by saying, 'We are satisfied with the type of wool the country will grow, instead of endeavouring to produce what the climate and soil continually fight against.' Today, those old stud ram pens are central to Peppin Park — a sheep and wool museum in the old school grounds by the Edward River in Deniliquin.

Right around the country some restored buildings become receptacles for a quantity of 'bygones' and memorabilia collected from around the neighbourhood: anything from old patent medicine bottles to ploughshares. Some of these 'folk museums' work well in their sensitive presentation of simple messages. The copper-miner's cottage at Moonta, South Australia, tucked away behind the old mines and preserved by the National Trust, evokes deep thought. But, by contrast, the last time I visited the Silverton Goal Museum, north-west of Broken Hill, it consisted of an indigestible collection of left-overs.

The grand season for Australian folk museums opened in the 1960s, partly as a result of the increased mobility given to hundreds and thousands of people by steadily increased post-war prosperity. Tourism had become a growth industry and folk museums, whether as private or public enterprises, developed as tourist attractions.

Our first folk museum of any consequence was established by a group of

This series of 'Pioneer Life' postage stamps was issued by Australia Post for general use on 15 November 1972 at the Swan Hill Pioneer Settlement. They depict vital elements of survival in pioneer society — water, food, shelter, transport, commerce and communications.

enthusiastic locals at Swan Hill on the Murray River. The promise of help from the Victorian government prompted them to buy the historic old paddle-steamer *Gem,* and relocate her in a specially dug pool near the river. This success sparked off a much larger venture. With advice from Melbourne architects and designers, they employed local craftspeople to reconstruct a pioneer township on land close to the *Gem.* They fossicked through the surrounding Mallee and Riverina country each side of the river for old buildings that dated from the pioneer period of the 1880s. If available these would be transported to the site, and then designed and assembled to be 'new' old buildings which would stand up to modern regulations, and thousands of visitors much better than the genuine antiquities.

The success of Swan Hill Pioneer Settlement, which emerged into a full-scale business, soon inspired other provincial towns in various states to follow suit. Ballarat built its impressive monument to its gold-mining heritage of the 1850s at Sovereign Hill. Gosford, north of Sydney recalled the first days at Sydney Cove by constructing and 'operating' a full-scale replica of the first settlement by Europeans at a 'look-alike' cove, accessible by tourists. Coal Creek in Gippsland tried the same thing but with a mining theme, as did Warrnambool on the Princes Highway in southern Victoria, with its Maritime Village Museum, not far from Tower Hill. Many more tourist attractions sprang up to compete for the tourist dollar, and motels named after the famous and infamous characters of each region proliferated around them.

Each bushranger received his memorial, a souvenir stopover near the appropriate busy highway, and some of the operators robbed the public far more effectively than Ned Kelly. The robbery has not been only in the

material sense; but in a way the more artificial of these folk museums have robbed us of our true heritage. When one actor 'flogs' another actor every hour on the hour, with fake blood and screams, for the delectation of those who have paid to watch the performance, it somehow takes away from the genuine anguish of those unfortunates who were forced to create Australia.

Perhaps the best one may hope for are activities such as panning for gold at the hugely successful Sovereign Hill. These certainly inspire the paying public to look deeper into our past and find a role to play in the preservation of our heritage.

Somewhere, some time, all this experimentation with telling and selling Australia's past may prompt the enquiring mind to sense the spiritual and emotional qualities of the Australian environment, and be astonished by what we have been missing.

The Final Part of the Model: 'How do aesthetic evaluations grow? How may poetry come to life in such a place as Australia?'

From stone-age to computers within two centuries represents a staggering rate of change, and yet that is what we have inflicted upon the continent of Australia. Perhaps it is amazing that within that short period we have not only impressed an imitation of European civilisation upon the land, but also developed some of our own traditions, and our own cultural and intellectual life. And yet this life of the spirit that is expressed for us by poets, writers and artists, also owes more to Europe than Australia. It was established by men and women who looked outwards, often nostalgically, instead of inwards, to the great cultural force which had long existed in the land.

Two Owls

The owl-man belonging in the dreamtime and created by Aborigines as a rock painting near Bala-Uru in Arnhem Land. Flying into the scene is the common screech owl of Europe to mark the coming of the white man.

Unless they were anthropologists, most people looked upon the spiritual and cultural life of the Aborigines as nothing more than paper manifestations.

Only recently have we begun to guess at all we have ignored. Now, when we would like to say sorry, and wish to incorporate it in a multi-faceted Australian culture, it is almost too late. It has become their own rallying point as the Aboriginal community strive painfully to restore themselves to some feeling of racial importance and dignity.

Aboriginal art and craft forms are perhaps more genuine than others because they spring from the way in which those sensitive people respond to the Australian environment. In song, dance, poetry, storytelling, decoration, marking and painting, they intermix easily with the environment. Such symbolic markings as the Dreamtime Owl, on rocks near Bala-Uru in Arnhem Land, were made to represent the legendary creatures which actually created their environment during the great Dreamtime.

Bill Harney, author

The journalist Douglas Lockwood came to know the spirit and dignity in his years editing and writing for newspapers from Darwin. He also knew the legendary Bill Harney better than most. In his foreword to Harney's book *Songs of the Songmen* he comments on the author: 'Perhaps the strangest aspect of his character concerned his ability to communicate easily with Aborigines. Few men got closer to them than he did. He was the only European I knew who could *think* like an Aboriginal. He was known by the tribesmen from Arnhem Land to the Gibson Desert, and his knowledge of them was remarkable. It was typical of Harney that he wasn't satisfied simply with being acquainted with members of a particular tribe; he had to know about their customs, their beliefs and ceremonials. He sat for hours and days and sometimes months with individuals and groups of elders to learn their story, and it was this kind of knowledge that made it possible for him to write…

'The Dreamtime heroes move: like phantoms deep
From distant ages – fleeting shapes that leap
Upon the black man's thoughts; each one recalls
Creek or river, springs or waterfalls
Within his tribal lands – from first to last
Recording deeds of heroes in the past.
Thus does the black man breathe again the air
Of Pakadringa, rain-man, and hid care;
Of Totem-heroes out of whom tribes leap
When Thunder-giants fought, and Lightnings swept
Throughout the land – to all a warning clear
To follow the path they trod, the way austere:
Of Dawn-folk springing from the misty past,
From Dreamtime nights, when life on earth was cast.

'Then the tribal saying: *'Mordja Amari Boaradja/Ngu Borngga Amari Mordja'*, which literally means *'Forgotten I lost dreaming/Country I left forgotten lost.'* So paraphrased — He who loses his dreaming is lost.'

AND NOW THERE IS CLANCY

Over the past century the ballad of *Clancy of the Overflow* has been almost continuously in print, and has attracted discussion, whether around campfires or in

A.B. 'Banjo' Paterson, poet

university lecture theatres. So much so that it is often difficult to think it is not a folk legend, nor a true story, but a set of verses invented by 'Banjo' Paterson who died only about seventy-five years ago. Clancy the invention has become the folk hero, and many arguments have been hammered out in the taverns of the country pubs and city bars on whether he is real or not, fact or fiction.

Our tradition, homemade here in Australia, holds that Australians should all be free and as independent as the spirit of Clancy galloping through the wide open spaces: vigorous, hearty, self-reliant, scornful of authority, and exempt of the niggling problems and spiritual erosion of city living. The fact that so many people seem to need him to be a real person is evidence of the profound and productive time Paterson spent preparing *Clancy of the Overflow* as a gift to inspire our inner lives as Australians.

Clancy may become our own substitute for the superhero tales we inherit from other lands and cultures, like King Arthur and Beowulf. Like them, Clancy both reflects and inspires our imagination. Intentionally, we believe, Paterson even structured the verses to show the deep schism between our outer and inner lives: the restrictions and realities of life in the 'Realtime', and the nostalgic yearning for escape to freedom somewhere beyond the horizon, if not actually the Dreamtime. In his ballad, the city folk yearn for the freedom of the wide open spaces, while the country folk suffer the reality of isolation and climate which will bring drought, flood or bushfire 'as the seasons come and go'.

Both have their visions of Clancy. Both cherish it in their spirit, but when they seek its reality he becomes more and more elusive. Looking for Clancy is like a search we make for something we urgently need but have misplaced. The harder we look for it the more frustrated we become. Perhaps it is better to leave him wrapped in his story, riding casually across the outback of Australian consciousness, singing his song, which echoes just beyond hearing, filling our dreams and hopes with all that we would like to be.

And in the place of lowing cattle,
I can hear the fiendish rattle
of the tramways and the buses
making hurry down the street.

And I somehow rather fancy
that I'd like to change with Clancy,
like to take a turn at droving
where the seasons come and go...

A.B. 'Banjo' Paterson, *Clancy of the Overflow,*

AXE FALL INTO SILENCE

During the occupation of Australia by white people, all urgently 'settling down', there has been little inclination for drawing up pacts, either with the environment or the Aborigines. The effect has been impact rather than pact, especially where the Aborigines are concerned, and it has existed ever since Captain Cook took a musket and fired it over the heads of two confused Aborigines at Botany Bay.

Daisy Bates, anthropologist

We cannot claim that the situation has not been explained to us, over and over again. In 1899, the London *Times* commissioned Daisy Bates to investigate charges of ill-treatment of Aborigines in Western Australia. She was able to show that the white man's impact was the result of ignorance and indifference, rather than of deliberate brutality, but the experience caused her to spend the rest of her life working for the Aborigines. In her book *The Passing of the Aborigines,* 1935, she argued that they had become trespassers in their own land. 'There is no more happy wandering in the interchange of hospitality. Sources of food supply slowly but surely disappeared, and they were sent away to unfamiliar places, compelled to change their mode of life, to clothe themselves in the attire of strangers, to eat foods unfitted for them, to live within walls.'

She believed that the Aborigines of the Australian mainland were doomed, just as the Tasmanian Aborigines were doomed to genocide by the axefall of white settlement.

Fortunately she was wrong, and the picture has radically changed since the days of Daisy Bates.

Early last century Herbert Basedow, an anthropologist, identified some of the many fundamental differences between black and white Australians. 'An Aboriginal's design may be crude,' he observed, 'but his imagination is, nevertheless, wonderful; we see the line, but he sees the life: we behold the image, he the form.'

The tragic consequences of our inability to understand the Aborigines is reflected in these words. We have tried to follow the undeviating line of white supremacy, while the true life and form of our country are seen most clearly by the Aborigines.

Queen Truganini, the last Tasmanian Aboriginal

What happened in Tasmania in the 1830s will never be repeated, and hardly mentioned here, except to say that by 1835 only a handful of Tasmanian Aborigines remained, and the last pure-blooded Tasmanian, Queen Truganini, died in 1876. A race of humans had been wiped off the face of the earth.

Some of the peculiarities in white attitudes towards Aborigines surfaced again in about 1960. Albert Namatjira, the famous Australian painter, had been made an honorary citizen of his own country and granted the right to drink alcohol, denied to others of his race. When he gave drinks to his family, he was imprisoned for breaking the law relating to Aborigines and alcohol. This sparked off such a fuss that the government eventually mounted a referendum in which white people voted on whether Aborigines should be allowed to drink or not. The majority was in favour.

One may question whether this was in the interests of the Aborigines. Alcohol is one of the dominant social diseases of the white community, and to make it freely available to Aborigines was hardly a forward step. Dr Charles Duguid, in his book *Doctor and the Aborigines,* points out that medically there are distinct differences between white and Aboriginal physiology. For countless generations the Aboriginal diet did not include any kind of yeast, but they are now exposed to introduced yeast concentrated in foods and alcohols.

The drinking question was only one of the implications of the referendum of the 1960s, which in effect made Aborigines citizens of Australia. This resulted in a great upsurge of Aboriginal insistence on taking their rightful place in the country of their ancestors. They wanted a share of the commonwealth wealth.

This insistence sparked the Land Rights movement. For years, Aborigines have been employed as stockmen on the outback pastoral stations, working for minuscule wages on the tribal territories of their ancestors. Little by little these station lands are being given back to them, and whether they work them as cattle runs or dig them up as mines on their own account is really of little consequence.

In the past, they had no say as to how the land taken from them and given or sold to white exploiters was to be handled. The new occupants could use it for stock, mine it for minerals or devastate it in any way they pleased. If there was a sacred site on the land, the Aborigines would not dare to object to its destruction. The churned-up landscapes of Coober Pedy, Andamooka and White Cliffs, where white men have torn up the ground with everything from pickaxes to bulldozers in their search for opal, and dug themselves dwelling-places below the ground, are typical of the European rape of the Aboriginal motherland.

Nowadays, the Aborigines are alert to attempted exploitation. If they stood alone they would perhaps have little chance of preventing it, but they are supported by many white groups, ranging from trade unions, legal interests and conservation groups. In such cases as Nabarlek in the north-west and Portland in the south-east, to say nothing of the uranium and aluminium projects in the far north, they voice their objections, knowing that they will be heard. As a response the organisations which exploit Australian mineral wealth have long ago learned that any mouth can be stuffed with money. From simple exploitation they moved into wheeling and dealing, so that Aboriginal communities now receive outright payments and mineral royalties for workings on their tribal lands.

Sadly, money fuels the juggernaut on its path towards chaos and wilderness. The Aborigines are no more capable of the prudent handling of earned income than is the average white person, and money simply taints them with the white person's materialist obsessions. The outback roads are littered with abandoned vehicles, gaudy plastic trash, empty cans, bottles and flagons for which the Aborigines have sold their birthright. Some sturdy individuals have managed to lead tribal groups back down traditional paths, but for all too many Aborigines, a share in the 'common wealth' has only been a push yet further away from the 'happy wandering' of their true heritage.

If there is any answer to this question it has not yet appeared. We can say sorry, but that is not enough. It is certainly not fitting for the non-Aborigine to preach to the Aborigines about their present lifestyle, when it is simply an imitation of our own.

Among the 'sturdy individuals' who have taken up the cause of helping Aborigines through this wilderness are white people like Ted Egan, a folklorist, singer and campaigner for justice. About his famous song *The Drover's Boy*, he writes:

'Old Aboriginal women with their timeless, stoical faces, sitting in the dust, smoking pipes, wearing broad-brimmed stockmen's hats, scan the horizon with the look that tells you they understand the big distance, and long dry stages.

'How did they spend their younger days? Ask them. You might get the throw-away reply, "I was a drover," or perhaps, "I was a stockman". And they'll tell you how they were better drovers than the men, better riders, better all round. They could find the water and they could track lost cattle. Better than the men.

'The story of these pioneering women of northern Australia has never been told before. They were dressed up as men, as "drover's boys", to get around the law that forbade the employment of Aboriginal women as drovers, and they made a huge, unacknowledged contribution to the cattle industry in Australia in a period that saw the biggest movements of cattle anywhere in the world.

Ted Egan, folklorist and campaigner

'Many of these women were victims of incidents like the "massacre in the west", mentioned in the song:

'...And he told of the massacre in the west,
Barest details, guess the rest,
Shoot the bucks, grab a gin,
Cut her hair, break her in...

'These women became virtual slaves, working without pay to move cattle all over Australia, often far from their own country. And yet in some cases, as this song suggests, genuine love between the "drover's boys" and their white bosses developed out of mutual dependence. Many of the children of such unions testify with affection how their parents truly loved one another. This is one of the many paradoxes of the Australian frontier.'

So when they build that
Stockman's Hall of Fame
And they talk about that
droving game,
Remember the girl who was
bed mate and guide,
Rode with the drover side by
side,
Watched the bullocks,
flayed the hide,
Faithful wife, never a bride,
Bred his sons for the cattle
runs.
Don't weep ...
for the Drover's Boy.
Don't mourn ...
for the Drover's Boy.
But don't forget ...
the Drover's Boy.

Ted Egan, *The Drover's Boy*

THE HUNTER'S CYCLE

Thus goes the cycle round and round;
The Hunter treads the same old ground.
He knows each pool, each hunting spot,
Where food is good, cool days and hot;
The bush – it is his living plot.

Bill Harney, 'The Hunter's Cycle', *Songs of the Songmen*

Hunting, shooting, trapping and fishing is legitimate provided that the wildlife population is preserved and controlled, so that those who kill animals for pleasure, protein or profit cannot exhaust the various species. In the home countries of early settlers, the 'game' had been most strictly controlled, and in fact a good many settlers or their parents, had been transported for 'poaching on the Squire's preserves'. But the notion of wildlife preservation caught on very slowly in Australia and was ignored or resisted by the majority. The Aborigines had hunted wildlife only for food and skins and made little impression on huge animal populations, and they mostly did so as nomads, never staying in the one place for long. White settlers hunted native 'predators' to rid themselves of competitors, because they saw the indigenous creatures as a threat to their sheep, cattle, grain crops and orchards.

They succeeded all too well in the case of the Tasmanian marsupial wolf or *Thylacine cynocephalus*, often known as the Tasmanian Tiger. By the time of white settlement the thylacine was almost extinct on the mainland, which might have given a clue to its vulnerability if anyone had been interested, but it was still common in Tasmania.

In the Tasmanian ecosystem the thylacine was a carnivore which hunted wallabies and birds on the forested hills and mountains. When settlement spread into such areas after the 'removal' of the Aborigines, the settlers saw the creatures as a threat to sheep. They hunted them ruthlessly until 1910, when remnants of the species were afflicted by some disease, possibly distemper introduced by European dogs. At last the only survivors were to be found in the rugged West Coast wilderness, but by the 1930s these too were thought to be extinct, although unconfirmed tracks and sightings continue to be reported.

Tasmanian tiger
(Thylacinus cynocephalus)

Even while the settlers were eliminating the indigenous Australian wildlife, they were introducing creatures which would do immense damage to Australian ecosystems. Rabbits have been the most destructive but there were sparrows, starlings, foxes and ferrets. Countless domestic cats are wild and have developed a race of feral cats which are like miniature tigers preying on small bush creatures. Wild pigs, descended from farm stock, cause great environmental damage in New South Wales and Queensland. Settlers took goats into the outback to provide themselves with milk, and their descendants roam the Flinders Ranges in huge flocks which are steadily defoliating those ancient hills. Camels, donkeys and brumbies, all descended from work animals taken into central Australia, compete with native animals for food and water.

Our attitude towards the creatures of the natural environment is still somewhat confused. We make many attempts at conservation but we still allow random culling by shooting or poisoning when better solutions to population control are known. We create reserves, but make them so easy of access that thoughtless drivers of powerful heavy vehicles and riders of trail bikes can cause more damage in one weekend than the land suffered during centuries of Aboriginal occupation.

Obviously we cannot put the clock back and it is impossible to reverse the situation which has made Australia one of the great primary-producing nations of the world. The axe of industrialised humans has fallen and broken the silence of our natural and blameless wilderness, and we can only hope for the future by establishing a wise environmental policy covering the entire population. But, to date, there is no proven overall authority competent to monitor the entire Australian environment, especially since environmental problems are exacerbated by political ignorance, factional interests and, of course, increased commercial pressures. Each becomes the subject of a separate conflict, with the media taking the attitude of 'let's you and him fight'.

The impossible ideal is the declaration of a moratorium for every kind of development and exploitation pending the establishment of an environmental impact pact — a master plan for Australia in which permissibility is the measure and sustainability the method.

Without something of that kind there seems to be no way to stop the destruction of evidence: the sweeping aside of facts and artefacts which might eventually tell us *what we do not know* about Australia...and that is a great deal.

For example, a fairly recent discovery was that of pre-Ice Age tools near the junction of the Gordon and Denison Rivers in south-western Tasmania. Until then, all evidence indicated that Aborigines settled there during the Ice Age, when the island was connected to the mainland, 6–10 000 years ago. New evidence might show that they, or somebody, were here much earlier. But in the 1970s the Gordon and Denison Rivers were the subject of a conflict as to whether they should be used for a hydroelectric scheme. If they had been, then the last chance of finding such evidence may have

High, windy ridges where one can sit upon a crag and look into the misty distance, and feel the movement of the world, and perhaps — if one can cease to think about oneself — hear the sound, the music of it.

A strange and unexpected music that one senses rather than hears.

Ellen Miller, *The Quiet Land*

The Crags of Capella above Lake Cygnus, Western Arthur Range, South Tasmanian wilderness

disappeared. Fortunately the energy scheme did not go ahead, but the conflict drew international attention to the issues surrounding nature and heritage conservation.

The earliest attempt to establish any kind of controlling body was the creation of the Australian Conservation Foundation in 1965. This was a non-profit organisation which aimed to promote the practice and understanding of conservation. It had a moral rather than a legal authority, and it did a great deal to make Australians aware of environmental issues.

In 1972, when Prince Philip was the President and Sir Garfield Barwick the Vice-President of the ACF, it became concerned with the battle to save Lake Pedder in southern Tasmania. Prince Philip promoted the view of scientists with this broad observation: 'In a pioneering society, the principal problems are to survive, to exploit and develop. The Lake Pedder case marks the end of Australia's pioneering days and it ushers in a new phase of conscious concern by all sectors of the community for a long-term future of the natural and human environment.'

Sir Garfield's hope was that 'the tragedy of Lake Pedder' would 'stimulate the establishment of proper land-use authorities throughout the Commonwealth…indeed hope that no other part of Australia will be pedderised, i.e. unnecessarily damaged as a result of inadequate consideration of all the values concerned and without adequate disclosure to the public of the contemplated use.'

Lake Pedder was not saved from change, but it probably marked a turning point in Australian conservation thinking. The Commonwealth did establish 'proper land-use authorities', including the National Parks and Wildlife Authority and the Great Barrier Reef Marine Park Authority, and they have some legal power. Ironically, Lake Pedder is developing a new beauty and a fresh ecology, far more accessible than in the past, and with proper care it will provide new aspects of natural wilderness for future visitors.

In the 1970s, during the Lake Pedder battle, the Victorian state government conducted a far-sighted scientific and social study of the Westernport Region, the 'Westernport Bay Environmental Study'. Professor Maurice Shapiro was to direct the Study, which was the responsibility of the Ministry for Conservation. In 1973, Shapiro wrote: 'Environmental studies allow us to introduce a new dimension in the planning process by bringing together economic, social and environmental considerations. This development is more significant than the mere introduction of a new element into the decision-making process. Potentially it involves the complete reorientation of technology towards conservation rather than exploitation. If this potential is realised, we will have true wisdom in our choices.'

In the past, studies were designed to determine the extent of environmental degradation, in order to establish a program for undoing the damage. In contrast, the Westernport Study was one of the first in the world to be set up *before* major development, in order to provide sound environmental planning.

The Westernport region covers some 3000 square kilometres around the shores of Westernport bay, a deepwater harbour about 56 kilometres from the centre of Melbourne. The region, including the bay, brings together a great many things that people value: natural beauty and abundant wildlife, clean sandy beaches and an atmosphere of serenity and peace. But it also has great potential as a deepwater port and industrial centre; so the question to be answered was — can we have all these things together? If we develop Westernport further, need it change the environment?

Illustrations from Westernportrait, published in 1973 to inform the public of the aims of the Western Port Bay Environmental Study.

When the study was completed and options for the future of Westernport were approved by government, industry, conservationists and the general public. Now, thirty years on, those options still form the frame for assessing the right path for development and conservation in that region.

Churchill Island covers only 57 hectares and is in Westernport Bay, reached by a causeway from Phillip Island. Churchill Island has a historic place in the foundation of Australia. Discovered in 1798 by George Bass, it was first settled by Lieutenant John Grant and a small party of guards and convicts in 1801. They planted wheat, corn, peas, rice, onions and potatoes and various fruit trees, in the first attempt to create a 'European' type of community in what is now Victoria.

They were unsuccessful and Grant moved to Tasmania with his convicts. They left behind what was probably the first building on the Australian mainland apart from the settlement around Sydney Cove: a wooden blockhouse. The signs of their settlement had long disappeared when private owners took over the island in 1856, but nevertheless Churchill Island is historic heritage – the documented site of the first European settlement in Victoria.

About the time the Westernport Bay Environment Study was in progress the Victorian Conservation Trust was set up by Sir Rupert Hamer, then Premier of Victoria. Its responsibility was to encourage and assist in the preservation of those areas of Victoria which are ecologically significant, of natural interest or beauty or of historical interest. Churchill Island certainly came within these categories and the VCT soon recommended that the Hamer Government should purchase it. The fact that the government did was a fine example of buying property for the public in the future, even if there were no definite plans for its use at the time.

Today, Churchill Island is a passive, well-managed place where visitors and wildlife are welcome. The Conservation Trust has refined its watch over Victoria's heritage and now has many such land holdings or 'covenanted' properties. Under the current title of 'Trust for Nature', a rare association has been forged — government and individuals have a pact that is central to the conservation ethic. And it is necessary to go to the far north of Queensland to realise how much has been achieved by quiet, often unknown individuals.

Overview of Churchill Island, Westernport Bay, Victoria

The great conservation issues appear to be the concern of government, industry, trade unions, local authorities and similar faceless groups, but in reality the vital issues are raised by dedicated individuals. They are the ones who keep the thread of conservation intact.

John Busst was one of those individuals. As a northern Queenslander, he made a lone stand in 1967 when the Ellison Reef within the Great Barrier Reef was mined for commercial limestone. The impact of this lone stand against industry and government, in the tiny Mining Warden's Court at Innisfail, rallied the conservation concern of the nation behind him.

Following that success, he became increasingly effective as a spokesman for the Reef. Perhaps he was helped by his close friendship with Prime Minister Harold Holt, who is supposed to have told him that the federal Government would take over the Barrier Reef if he lost the battle to stop coral mining, or if the Queensland government endangered the Reef's future.

Eric McIllrie of Dunk Island is another Queenslander who fought for the Reef, with the backing of John Gorton after Holt was lost at sea.

The spectre of John Busst still haunts those who try to interfere with the mainland coastline from south of Heron Island off Gladstone to the north of Lizard Island and much further. His work is continued by the many individuals along the coast, into the rainforests and throughout the islands of the Reef area who strive devotedly for 'the preservation of the wild'. They include Margaret and Arthur Thoresbourne, who lived as nature's tenants in a tropical garden on Meunga Creek near Cardwell. Their efforts are known and respected by all the conservation-minded community on the seaboard of northern Queensland. The great walking track that traverses the wilderness of Hinchinbrook Island is named after them.

Arthur and Margaret Thoresborne with a friend counting Torres Strait pigeons on the Brooke Islands, Far North Queensland

Much of their early work was devoted to the study of Torres Strait pigeons, a threatened species which returns late every winter to feed and breed in the rainforest of the Brooke Islands, seaward of Hinchinbrook Island. In the late sixties and through the seventies Arthur and Margaret sailed there each year to monitor the arriving flocks, check on their survival rates and harass the soulless who slaughtered the pigeons for 'sport'.

After Arthur Thoresbourne died, Margaret and her devoted followers continued to harass the developers who contrived to make financial gain from the environs of Hinchinbrook Island and its channel that separates it from the farming lands of the mainland. The Port Hinchinbrook development of a marina and tourist sub-division for houses and resorts overlooking the channel and the island was held up for years by her efforts and patience.

Unfortunately most Australians, by the nature of their lifestyle, never have the opportunity to learn this kind of patience. It is easier for them to take nature's bounty and human greed for granted, as something which appears extravagantly packaged on supermarket shelves or flows out of a nozzle at the service station.

Only slowly have we been forced to recognise, primarily by the efforts of the dedicated individuals and small groups of the conservation movements, that important non-renewable resources can be exhausted and that even self-generating resources, such as lobster populations and coral reefs, can be depleted by over-exploitation or careless damage to the environment.

Fortunately there is a World Conservation Strategy that can help everyone who is concerned about our world. The Strategy says that our way of life can only be sustained by conserving the living resources — the plants and animals — on which our lives depend, and by permitting development with conservation. It promises that, through conservation, ways can be found to repair the damage that has already been made to our land, air and water without upsetting the demands of technology and progress.

To celebrate the adoption by Australia of the World Conservation strategy in 1986 the South Australian Department of Environment and Planning published a set of educational posters (reproduced here). After that beginning there has been little substantial follow-up. Imaginative ideas have emerged, then gone away, seemingly without trace — notions like the creation of a National Herbarium as a safe place to store and preserve our heritage of plant collections and, on a much grander scale, the establishment of a Commonwealth EcoBank to fund enough money to create and maintain environmental research, education and wilderness experiences.

In 1972 a treaty known as the World Heritage Convention was adopted by UNESCO. Its goal was to recruit the world community in identifying cultural and natural properties of 'outstanding natural value'. There are now about 800 sites incorporated in the World Heritage Sites program that aims to keep track of and control the Earth's lasting treasures. Nations lobby hard to get their buildings, wilderness and historic places on this list, and the listing approval brings prestige, tourist income, public awareness and national

Poster series prepared by the South Australian Department of Environment and Planning in 1986 to promote the goals of the World Conservation Strategy as it impacts on the South Australian environment.

Forests *are like families, spanning generations, fulfilling our lives and being replaced in time.*

The big difference is that forests are felled while in their prime to build houses and farms and cities.

Wildlife *is like a secret, hidden in the recesses of our mind from the view of other people.*

The big difference is that a secret is ours to give away but wildlife is all the earth's to share.

National Parks *are like promises, symbols of our values and trust, providing hope and security.*

The big difference is that parks are promised to the unborn and need us as their guardians.

Trees *are like umbrellas, raised to shelter us from the sun, wind and rain.*

The big difference is that trees spread their canopies to protect birds and animals and land.

Rivers *are like conversations, long and short, fresh and dull, restful and challenging.*

The big difference is that rivers are water and fish, weeds and birds, not words.

COASTAL DUNES
are like ideas,
continually being created and shaped,
sometimes vanishing without trace.
The big difference is
that dunes and beaches are not formed by the mind
but by the sea and wind and time.
NATIONAL CONSERVATION STRATEGY FOR AUSTRALIA
RESOURCES FOR LIFE
DEPARTMENT OF ENVIRONMENT AND PLANNING, SOUTH AUSTRALIA

ENERGY
is like money,
often slipping through our fingers
without trace or gain.
The big difference is
that nature's energy cannot be lost
and the sun and winds and tides are always with us.
NATIONAL CONSERVATION STRATEGY FOR AUSTRALIA
RESOURCES FOR LIFE
DEPARTMENT OF MINES AND ENERGY, SOUTH AUSTRALIA

MANGROVES
are like nurseries,
nurturing and sheltering the young,
preparing them for the world beyond.
The big difference is
that we tend nurseries with care and devotion
but mangroves are left to fend for themselves.
NATIONAL CONSERVATION STRATEGY FOR AUSTRALIA
RESOURCES FOR LIFE
DEPARTMENT OF ENVIRONMENT AND PLANNING, SOUTH AUSTRALIA

Coastal Dunes are like ideas continually being created and shaped, sometimes vanishing without trace.

The big difference is that dunes and beaches are not formed in the mind but by the sea and wind and time.

Mangroves are like nurseries, nurturing and sheltering the young, preparing them for the world beyond.

The big difference is that we tend nurseries with care and devotion but mangroves are left to fend for themselves.

Energy is like money, often slipping through fingers without trace or gain.

The big difference is that nature's energy cannot be lost, and the sun and winds and tides are always with us.

pride. Australia has been successful in having a range of natural sites listed, but as yet no cultural precincts. The sites range far and wide; from the Great Barrier Reef, Kakadu National Park, the Tasmanian wilderness, islands like Lord Howe and Macquarie, and places where our ancient beginnings are visible. In time our cultural sites may be recognised and listed too. Places like the Sydney Opera House, the Melbourne Cricket Ground and even Flemington Racecourse.

The notion that Australia has a heritage worthy of protection and preservation is comparatively recent. The growth of a national consciousness was confined to historical societies and some local museums that struggled to survive in an unsympathetic society. At a seminar in 1967 on 'Historic Preservation in Australia', architect Robin Boyd spoke of the prevailing attitudes: 'To some other people, thoroughly good citizens to all appearances, the cause of preservation of old buildings and ancient trees and other items of national heritage is not self-explanatory. They cannot understand it at all. Or rather, they can explain it, but only as sentimentality with no meaning to the modern world. Often they see it as anti-progressive, reactionary; quite often it irritates them.'

Progress in changing these attitudes has been slow, and even today, forty years on, the old irritations surface, and battles are fought once again. With the huge re-development of the Members Stand and facilities of the Melbourne Cricket Ground there were many that felt it worthwhile to argue for no change to the uncomfortable seats and winter winds that was the heritage to be endured in our great cathedral of sport.

FOLIO TWO

MARKING TIME

Why do we need to preserve our heritage? and How can we choose what parts to conserve and what to destroy?

One may as well ask, 'Why keep a family photograph album?' The answer is that one of the most important needs of the human spirit is recognition of identity.

Each of us needs to know, for our mental and spiritual stability, exactly where we fit into the scheme of things. It is a need as fundamental as the Aboriginal tribesperson's need to know their territorial boundaries, and all the shapes and signs and sounds of the land which make them feel at home. If we destroy all that which links us to our past, we are doomed to wander in a wilderness where nothing is permanent because only figures on the balance sheet have value.

The answer to the other question — how to choose what we conserve — is more complex. We know the *cost* of almost everything, but we have not yet devised real ways to estimate *value*. The cost of a full-grown tree, as timber, may be a few hundred dollars, but its value to humankind is measureless. Quite apart from the emotional aspects, we know that trees are a vital part of our life-support system. An easy way to species suffocation is to cut them all down.

It is equally difficult to assess the value of an old building, a beautiful stretch of coastline or even a ghost town. As the title of the National Trust indicates, such things are held in trust to use for our own appreciation and to pass on to future generations. We cannot even guess at their value in the future. We are all familiar with the progress of artworks, once sold for no more than would buy their creator a good meal, which are now exchanged for millions of dollars. Similarly, many buildings once scheduled for demolition are now cherished features of our community.

We know for certain that once anything has been destroyed it can never be restored. We must study each item, not as a part of a profit-and-loss account but to assess the satisfaction which it will bestow upon our descendants. They may, by then, have come to understand that *enough* is the only bottom line — not profit or loss.

There is a parallel between the transfer of our historic heritage from one generation to another, and the transfer of real estate from one person to another. In real estate, the vendor must prove their title to the property. A deed is drawn up describing the property, and the purchaser accepts a valuation. But there is one vital difference. In the transfer of our heritage, those who will accept it at our valuation are still unborn and must be represented by trustees.

We, despite our problems and our immaturity, are those trustees.

THE DUTCH BARNS

'Brickendon', Longford, Tasmania

These three great 'Dutch-style' barns form an impressive trio and dominate other farm buildings in a setting which looks almost English. 'Brickendon', which has been in the Archer family since the 1820s, is still an active property and these barns are used to shelter farm machinery more suited to modern grazing and cropping.

'LINCOLN'S PLACE'
Bellarine Peninsula, Victoria

Long before the development of refrigerated milk transport, this old cheese factory processed the milk from surrounding dairy farms on the fertile Bellarine Peninsula on Port Phillip Bay. The factory supplied cheese and butter to colonial Geelong and Melbourne. It is even said that in 1850 a cheese from this modest farm-factory won a prize at the Great Exhibition in London.

Today the building is in ruins and the farm paddocks are subdivided into modern home blocks for commuters who work in Geelong and beyond.

Pioneer Port

Robetown, South Australia, circa 1865

In the 1850s Robetown in the south-east of South Australia was a thriving port second only to Port Adelaide. Chinese gold diggers landed there to avoid Victoria's poll tax, and made the 'long march' to the goldfields. Sailing ships loaded horses for India and wool for Europe. Laden bullock drays off-loaded in the Royal Circus, near the old Telegraph Station.

Telegraph Station and Post Office

Robe, South Australia

Charles Todd, who was the father of the Overland Telegraph from Darwin to Adelaide, visited the Robe District in 1856. He recommended that a 'magnetic station' be installed at Robe as a vital link in the Adelaide–Melbourne telegraph system. This station was opened in 1858, and although the Morse telegraph facility closed many years ago, the Post Office operated until the mid 1960s.

Marwick's Trading Store

York, Western Australia

A Kalgoorlie gold prospector in 1893 told in his diary of the slogging march from Perth to the new goldfields: 'A good number of men I met say there is no tucker on the fields and that half the men are starving … we have had nothing but tinned meats since we left York.'

Marwick's trading store is on the old goldfield road at York, where it established a main calling place, a kind of early supermarket, for the coaches and drays, footsloggers and barrow-pushers on their way to find fortune at Coolgardie and Kalgoorlie.

Imperial Hotel

Ravenswood, near Charters Towers, Queensland

During the 1870s gold discoveries in north Queensland established new townships. Most no longer exist. At Ravenswood, inland from Townsville, gold was discovered in 1868, and within a few years it was a booming town with thirteen hotels and a railway. Now only the Railway Hotel and the Imperial remain for travellers to photograph and wonder at.

GOVERNMENT HOUSE
Hobart, Tasmania

' ... Softness of sandstone and grace of English trees. Old homes at peace in unexpected places ... What kind of life could they establish, knowing no other, their roots nourished by a thousand years of English village life? Cottages and shops huddled around church and graveyard, a manor in the background. Manor and church, security in the now and the hereafter, tenuous, but all they had ever known or imagined it was possible to know. So — cottages, shops, the church, the Government House equivalent of the manor ...'

The Quiet Land, *Ellen Miller, 1977*

THE BARWON HEADS GOLF CLUB

Victoria

Clubs are everywhere, and most people 'belong' to some kind of club as a member because they share interests and beliefs with other members. There are social clubs, service clubs, sporting clubs, indeed clubs for everything. In a sense, clubs are the thread that sews together the fabric of our society, and allows the adding and undoing as we judge what we should keep and what we can discard in our lives. The grand Barwon Heads Golf Club is an institution founded on a Scottish heritage and transplanted a hundred years ago to the sands of the mouth of the Barwon River near Geelong. The clubhouse has been re-styled many times over the years but retains its essence of 'club' tradition and continues to thrive.

Como House, *South Yarra, Victoria*

Como, built in the 1850s on rising land south of the Yarra River, is a reminder of grand colonial Victoria following the gold rush.

THE HERITAGE WATCH

The term 'heritage,' like conservation and environment, has only come into common use in the past two decades. Before then it would have been difficult to find an Australian who could have defined it in relation to our country. The vast majority did not care, and they took little notice when the houses and other buildings erected during Australia's colonial era were swept away by bulldozing demolishers or left to the vandals, because few people could see them as anything but unimportant relics standing in the way of progress.

Slowly the character of the country was being changed. Architectural self-advertisements in steel, glass and plastic ensured that every main city centre in Australia began to look like downtown Tokyo or Toronto. The new buildings were constructed in defiance of all the warnings about energy-resource depletion, and now, with their dependence upon electricity for air-conditioning, lighting, data processing, communication and elevator operation, they have become energy-guzzling monsters.

Fortunately, a small but determined resistance movement had begun to be active back then, and was fighting a 'guerrilla' war against this trend. They comprised the various State branches of the National Trust of Australia. The first was that of New South Wales, formed in 1945, and the sequence thereafter was South Australia, Victoria, Western Australia, Tasmania, Queensland and the Northern Territory. A co-ordinating body, the Australian Council of National Trusts, was created in 1965, and within twenty years the total membership was about 60 000. The individual branches are independent organisation but they all have the same aims.

These are to preserve and protect for the public, often by acquisition, any lands, buildings or other features which may have national, historical, aesthetic, architectural, scientific or cultural interest; to conserve natural features, scenic landscapes and wildlife; and to foster public knowledge, appreciation and enjoyment of these things.

Supported by this very wide brief, and often 'unpaid' members, the National Trust has won battles which often seemed hopeless. Like other movements it has developed from very small groups of dedicated workers. The Victorian Branch provides a characteristic case-history. It was formed in 1956 by a few dozen people to combat the steady depletion of the State's heritage, and they soon found a rallying point in the threatened destruction of the splendid old mansion called 'Como' in South Yarra. With only a few pounds in their treasury, the Trust members set themselves to raise enough money to buy the house and garden. By 1966 they had raised $325 000 to buy, restore and develop the property, and had attracted a good many new members.

Other states have had similar successes. But a most important function of the Trust is as propagandist and educator, lobbyist and 'gadfly'. Each time they succeeded in saving some property from the developers, or simply from inevitable decay, it was a propaganda victory as well as an actual victory. Success breeds success, and the effect of one triumph after another was to make more and more people believe, whether they were Trust members or not, that they could slow down the juggernaut of progress when it threatened the national heritage. But in recent years, slowing down progress has not been enough for some Trust members. On occasions they have sought to actually halt progress and have met with unshakeable opposition.

A symbol of preservation — the clock tower of the Olderfleet Building in Collins Street, Melbourne, built in 1889.

The Trust's victories helped to educate people to look at buildings with new eyes: to wonder whether an old building *had* to be replaced by modern filing-cabinet units and apartment houses, and to realise that age has a beauty and individuality which is not easily matched by the new.

It cannot be that every old building merits preservation, regardless of changes in communal needs, and it would be wrong to imply that the whole community should be preserved like a fly in amber without the opportunity for these changes and developments which are integrally human. Some marchers under the heritage banner may have their own emotional reasons for resisting development, rather than a genuine desire to conserve the natural and man-made places.

It must also be observed that some of the 'filing cabinets' have risen not because they were needed, but simply because a building or finance company saw them as investment opportunities. Such companies have thrust them skyward for no better reasons than those advanced by accountants and economists, notoriously fallible prophets, and have often gone bankrupt in the process.

The need, as always, is for a commonsense appraisal of the *genuine* development needs of the whole community, which would have to include such radical procedure as actually talking to ordinary people, instead of relying on statistics and computer models.

Fortunately there are heartening signs that some local authorities and councils are heeding the groundswell of public opinion set in motion by such bodies as the National Trust and the Trust for Nature, and that they too have been 'educated' into balancing the real needs of development against the proper conservation of our heritage and natural surroundings.

THE FIRST KANGAROO
In 1770 Sir Joseph Banks first described the kangaroo, and many thought it had to be an invention. He wrote, 'Nothing certainly that I have seen at all resembles him'.

THE BUNYIP
Long before European settlement the Aborigines believed in the existence of some dark creature of monstrous size that lived in the swamps, lagoons and billabongs of their tribal lands.

THE MAHOGANY SHIP
Legend has it that a Portuguese ship built of mahogany lies hidden in the sandhills near Warrnambool in southern Victoria. The wreck was seen by settlers between 1836 and 1880, when it disappeared.

LASSETER'S LOST REEF
For more than seventy years people have been fascinated by the story of Lewis Lasseter's lost reef. Does the fabulous quartz reef of gold that he claims to have found really exist, and if so where is it? Lasseter died of thirst in the desert in January 1931, close to where he claimed his discovery to be. That was west of Mt Olga at roughly 25°S and 130°W.

TRASH OR TREASURE?

The great problem faced by every student of the past is that of separating the harvest of truth, out of a crop which proliferates with all the weed of myth and legend. Human beings are natural storytellers and they like to add plenty of embroidery to a plain and factual tale, and so we have inherited many wonderful myths, legends and tall stories. We weave, pass on and re-weave such yarns as those about Portuguese 'mahogany ships', bunyips, Lasseter's Lost Reef and the mystery of the 'Geelong keys'. Sometimes such stories have a slight basis in fact, but very often we pass them on simply because they are colourful yarns and we all like entertainment and mystery.

It is important for us to know the truth about history without the exaggerated interpretation, but historic truth and fantasy are often so closely interwoven that it is increasingly difficult to separate them.

An example of this is the story of the beginnings of the Australian wool industry. If you ask the average well-informed person, 'Who founded the Australian wool industry?' the answer will almost certainly be 'John Macarthur'. And yet the truth seems to be otherwise. Governor King was actually instrumental in sending the first merino sheep to New South Wales, to replace the scrubby goat-like creatures already in the colony, and John Macarthur's young wife Elizabeth almost certainly was the first person who nurtured and developed the first merino flock.

It is however natural, given the nature of patriarchal society, that John should have received all the credit while Elizabeth, in all probability, did most of the work. John's portrait appeared once on our two-dollar notes; whereas our 'average well-informed person' might find difficulty in defining

Elizabeth Macarthur, pioneer

Elizabeth's place in the scheme of things. She may have preferred that it should be so, because she never failed her husband in love and loyalty, even when it was obvious he was verging on insanity.

The doubts now raised in our own times, two hundred years later, as to whether John Macarthur actually was the 'father' of the industry, and whether Elizabeth should be 'mother', highlight our problems in separating fact from legend. We cannot choose what we receive as our heritage, in both tangible and intangible states, but we can at least study this vast and varied inheritance very carefully, in order to pass on those parts of it which have a definable significance.

This immense task has been begun, broadly speaking, by the Australian Heritage Commission, which is a statutory authority established under an act of the federal parliament in 1975. The function of the commission is to prepare and maintain a register of the National Estate. The register is an inventory of the natural and cultural heritage of Australia, ranging from national parks to Aboriginal sacred sites and from historic buildings to wildlife habitats, but excluding 'portable' objects such as paintings and artefacts.

So far, the register contains over 13 000 entries and it expands continuously, according to suggestions provided by the commissioners and accepted as worthy of inclusion. Obviously the commission works along rather the same lines as the National Trust, but there is no conflict between them. In fact they complement each other, especially since the commission is able to cover aspects of the natural heritage such as forests, swamplands and Aboriginal sites, outside the orbit of the National Trust.

The Australian Heritage Commission does not have a subscribing membership and thus lacks a dedicated corps of fundraisers of the type which enables the National Trust to buy and renovate buildings or other features. The strength of the AHC lies principally in 'moral suasion', in the sense that it can provide the nation with details, recommendations and reasoned arguments to support public opinion concerning its heritage.

Despite the work being done by the Heritage Commission and by various other local groups and bodies, a great deal of our heritage still slips through the net. Historical objects come down to us in so many different guises…artless artefacts and other odds and ends which have somehow survived, or products of human ingenuity in making life bearable in hostile surroundings, such as the pioneer windmill shown here, or much more peripheral items like those mentioned earlier that hang on my studio wall.

This pioneer windmill is a celebration of the ingenuity of early Mallee farmers, and is displayed in the Swan Hill Pioneer Settlement on the Murray River.

Imagination and reality are so inextricably interwoven in all the thought processes of mankind, as the Aborigine understands so well in what they call the 'Dreaming', that the significance of any inherited item depends entirely upon how we interpret it or how it may be seen by our successors. If any group devoted to the preservation of our past was given a choice between keeping Field-Marshal Sir Thomas Blamey's full-dress uniform or an old Anzac water-bottle, they would almost inevitably plump for the former. And yet the water-bottle which quenched the thirst of a Digger who carried it from Gallipoli to St Quentin would be more truly representative of the terror, courage, anguish and triumph of Australians at war. It all depends upon your imagination.

This realisation is prompting many people and authorities to move from the tangible to the intangible in their preservation of our heritage, in what might be described as expanded historical research. The preservation of a fine old home or of a pioneer's cottage is a laudable endeavour but it has little real significance unless we can really understand its meaning, not only to us and our successors but specifically to those who actually lived in such places.

Our view of history is in a sense 'sanitised' by costume dramas on television or the wide screen and even by historic reconstructions such as the Swan Hill Pioneer Settlement. They enable us to *see* what life was like for our predecessors but they can only point the way into the most difficult exercise of all: that of *feeling* what life was like for the people of the past. We tend to look upon their lifestyles as 'quaint', whereas they would not have seen themselves as 'quaint' at all, and whenever we gaze into the past it is as though we are glimpsing a kind of fragmentary piece of theatre acted especially for our benefit and entertainment.

To see it all in perspective we need to imagine the people of a century in the future trying to guess what life was like in our time. No doubt they too will see us as 'quaint', because they will not be able to imagine the multitudinous details of our daily lives, and will have to guess at them only from surviving records and artefacts.

For example, it is probable that life will have changed to such an extent that a person living in the first decade of the twenty-second century will find it impossible to imagine a city obscured by a blanket of smog, or a world saturated by digital and electronic communication tricks.

The National Library of Australia archive is one example of current efforts to recapture 'living history' for access by a curious public of today and for the future.

In 1980 the Government of the Northern Territory placed the following advertisement: 'The Northern Territory Government is presently conducting an Oral History Programme for the purpose of collecting tales, reminiscences

The archives of CSIRO contain photographs of their invention — the first light-weight air warning radar installation (LW/AW Mk1), near Darwin, 1946.

and recollection of former Territorians and also for the purpose of recovering and placing into the Northern Territory archives, photographs and documents of historical value. Any former Territorian who would like to be interviewed or who has material which they would be happy to see returned to the Territory should contact…'

This seems to be the first time that a state or territorial government has attempted to 'harvest its history' from documents which may be scattered all over the nation, and specifically to recapture the spoken memories of pioneers before it is too late.

The archives of various state libraries, and especially the National Library of Australia in Canberra, are eager now to collect material from 'vanishing Australians', and they employ manuscript officers whose task is to ferret out such data so that it may be preserved. An enormous quantity has disappeared, usually because it was unthinkingly destroyed after the death of a relation, but there is an increasing awareness of its value.

A CASE HISTORY — THE HERITAGE OF WOOL

Woolsheds such as the one at Tolarno are a vital link with Australian social history and heritage. Tolarno represents a great 'factory in the bush'. In the prosperous period from about 1860 to 1910 these outback buildings processed the wool in the first stages of its conversion from the sheep's back to textile suppliers around the world.

The Tolarno woolshed marks the site of a significant incident, still somewhat shrouded in mystery. It stands on the banks of the Darling River, north

of Pooncarrie, and although the shed is privately owned and operated, the public road passes between it and the river. If you follow the road south when the river is low you may be able to see the few remaining ribs of the river steamer *Rodney*. During the great shearers' strike of 1894, the steamer was bringing a 'board' of about forty-five 'scab' or non-union shearers upstream to work in the Tolarno woolshed. Striking shearers gathered at Moorara Station, downstream from Tolarno, and determined to stop them. When the steamer tied up for the night a band of strikers swarmed aboard, tossed the scabs into the river and burnt the *Rodney* to the waterline.

This was one of the incidents which sharpened the focus on the great strike in which the shearers and other workers fought for tolerable conditions. It helped to give unions in Australia a confident foundation and a rich heritage.

Today the Tolarno woolshed is in sound condition and only altered slightly to meet with modern shearing practice. It is a fine example of the 'timber-and-iron' tradition of the classic Australian woolshed.

Dennys Lascelles woolstore, Geelong waterfront, built in the late 1800s, is now the site of the National Wool Museum.

Far from the 'outback' of the Darling River and the dramatic times of long ago, the modern city of Geelong, on Corio Bay in Victoria, makes an interesting new chapter in the heritage of wool in Australia.

Geelong traces its history back to 1835, when John Batman 'bought' 100 000 acres of land on Port Phillip Bay from the local Aborigines. The township was proclaimed in 1838, was depopulated by the Ballarat and Bendigo gold rushes in 1851 and became a city in 1910. Its historical significance has been as a point of entry and exit for the rich wool-growing regions of Victoria's Western District, and as the place where wool is sold and processed.

In the early 1980s the planning authority for the city, the Geelong Regional Commission, gathered experts together to 'dream' about solving the problem of holding onto the past and making way for the future of retail development in the Geelong central business district.

Many of the old woolstores along the waterfront of Corio Bay were unused and in danger of being demolished. One store, Strachans, was turned into a supermarket and carpark, and Dennys' bluestone building was imaginatively re-developed into a National Wool and Textile Centre. The plan was to create a meeting and rally point for the ever-changing fortunes of the wool industry. It was to be a place where a visitor could see textiles being processed into garments; where wool fashions and trends could be displayed; and where the industry administrators, researchers and operatives could meet and make plans. The dream did not eventuate, and today the building attracts a few visitors to the National Wool Museum. However, nearly next door the Dalgety's woolstore building has been wonderfully recycled for Deakin University. Central to the university complex is 'Costa Hall', a performance

This large painting hangs in the National Wool Museum in Geelong. It depicts the historic contribution made by Geelong and its Western District to the wool industry. The three sectors are: pastoral and woolgrowing; handling and selling wool; and textile and fibre processing.

An irony is that Geelong has led the world in finding alternative techniques to improve that which it did best in the traditional way. These innovations have forced industry management to retreat from Geelong and leave empty stores on the waterfront and mills on the Barwon River.

auditorium that is becoming recognised by musicians worldwide. The almost entirely timber interior gives a sound quality that has become the envy of much more prominent and expensive venues.

So far, over a quarter of a century on, the compromise of welding the past and the now seems to have been achieved along the lines of that made for some of the great port warehouse precincts of London, Liverpool and Baltimore. New methods of cargo handling, especially containerisation, have bypassed some of the dock areas in those seafaring cities and made them

redundant. The old places have been converted for a variety of uses, including tertiary education, apartments, marinas and restaurants.

The danger is that of sudden and perhaps irrational decisions of the type which are easily made when both parties to an argument become wearied of the whole affair, or when one side seizes a sudden opportunity to conquer. The tearing down of the Belvedere Hotel in Brisbane in the dead of night some years ago is one example of this, although the battle to preserve some Sydney waterfront areas has had more success with the help of union 'green bans'.

SOMETHING OUT OF NOTHING

Slowly immigrants from Europe and elsewhere are changing Australia from a nation modelled entirely on British customs to a multicultural community which, although British in its systems of law and government, is being enhanced in countless ways by the influence of other lifestyles.

Every country needs its own kind of 'folk' theatre, which should spring out of a genuine folk tradition. In many parts of the world, folk-dancing fulfils this need for celebration and often has its roots in religious ceremonies or in the representation of such basics as sex, birth, harvest and death. In Britain the daily Changing of the Guard, which once had a military purpose, has now become a kind of folk theatre. The countries of southern Europe find emotional release in the great public ceremonies of the Roman Catholic Church. Oberammergau has its annual Passion Play, which was created, like the passion plays of medieval France and Britain, as an act of thanksgiving and a warning to wantons.

In Australia it would seem that the corroboree ceremonies by Aboriginal tribespeople are our nearest approach to this kind of living theatre. But looking more closely, white Australians are forging and quietly acting out their own unique kind of folk theatre in the form of the annual country agricultural show. Without a better understanding of its origins, values and potentials, we may lose this tradition as life on the land becomes an impossible struggle.

The origins of the agricultural show are somewhat obscure but no doubt they were adapted from fairs and markets and similar functions in country towns in England. In any case they became eagerly anticipated annual events in country Australia when they developed into a kind of mini-Olympics for pastoral skills and achievements.

Competitiveness is another basic human instinct, and the shows provided arenas in which farming men and women could demonstrate skills in stock management, horsemanship, superb livestock, fruits and grain. Women competed for dominance in such country crafts as quilting, preserving and jam and cake making. Often shows included trials of strength and skill which could range from boomerang throwing to sheepdog trials, from sheepshearing contests to footracing, horseracing, buckjumping and tug-o'-war battles. In the more heavily forested regions the 'woodchop' became one of the most eagerly contested displays.

Folk theatre, Australian style. These competitors at a country show in north-west Tasmania, dressed in traditional uniforms of singlets and 'creams', maintain a genuinely Australian tradition from our pioneering past.

Woodchopping is true folk theatre because it derives directly from everyday folk activity. Over the years it has assembled a ritual of its own. Nobody knows how this ritual developed: its rules do not seem to be written down: it is sufficient to know that it is there. To depart from it would somehow depreciate the meaning of this celebration of strength and expertise. Burly 'actors' in white singlets and spotless creams follow correct postures and rhythms for 'standing block', 'underhand' and 'tree-felling', amidst flying wood chips and flashing axes. There is no script apart from the mute fury of action and the hiss of expelled energy.

Woodchopping has become a familiar event in our country shows, and in some of the city shows which are their large-scale copies. It is as typical at showtime, from the Royal Easter Show in Sydney to the November show at Currie on King Island, as fairy floss, Punch and Judy performances, and boxing troupes. But it is the one that is our own folk ritual — a survivor from the days when people were proud to demonstrate, in public, the skills which they developed during the gruelling labour of everyday life. Such skills have been pushed aside by technologies which put the chainsaw into the hands of a timberman instead of an axe.

The survival of woodchopping as a piece of folk theatre is important to us because it is one of the few remaining links with the time of the pioneers,

when people worked directly with their hands instead of indirectly through machines.

Together with sheepdog trials and demonstrations of sheepshearing, woodchopping relates back to a comment by C.E.W. Bean, the journalist 'recorder' mentioned earlier. In 1909 he wrote: 'It is still the quality of the Australian that he can make something out of nothing...he has had to do without the best things, because they do not exist here. So he has made the next best do; and, even when these are not at hand, he has manufactured them out of things which one would have thought it impossible to turn to any use at all. He has done it for so long that it has become much more than an art. It has long since become a part of his character, the most valuable part of it.'

Colonial crafts. There is a precious value of pleasure and beauty in items of colonial craftmanship that express our talents for improvisation. Here are a few examples: a hay fork and a clothes peg showing the influences of European peasant crafts; a forged iron wheel; and a drover's tripod and camp oven. The Coolgardie Safe used wet hessian sides to keep the contents of the safe cool under colonial farm verandahs; and this Windsor chair was fashioned from eucalypt hardwood with saw, adze and knife. The windlass barrel was invented by goldminers for raising heavy loaded buckets of ore from mine-shafts. And Ned Kelly's helmet was fashioned from plough shares in the late 1870s by the notorious bushranger.

There is a precious value of pleasure and beauty in items made 'out of nothing' — colonial craftsmanship. They express our talents for improvisation, our ability to make do with whatever materials are close to hand. Those talents and abilities may have become obscured by modern mass-production and automation, but they are still there whenever we need them, the 'most valuable part' of our character. Our deep need to be able to create for ourselves is shown by the increasing popularity of 'hobby crafts', from pottery making to woodturning, knitting and macramé, and it is interesting that the products of such craftsmanship are often modelled upon the simple craft productions of past times. In those days they were simple out of necessity, but their very simplicity possessed the appeal which springs out of most items created for a basic human need rather than for 'conspicuous consumption'.

Most of the original craft workers are gone, but their implements, utensils and folk art remain as a testament to their character, which is part of our inheritance. Many remnants of their work were tossed aside as useless when machine-made artefacts took their place, but we must be sure to cherish whatever remains.

We do not have to look far for encouragement. Just under thirty years ago in Victoria, the Premier, Sir Rupert Hamer, was encouraged to start the Victorian Tapestry Workshop. With advice from sculptor Lenton Parr, visionary Dame Elisabeth Murdoch and Sue Walker, the first director, a building was bought in South Melbourne and wool tapestries began to be woven. Artists were commissioned to adapt their paintings to be interpreted by weavers who employed age-old techniques for transferring design to frame, and weaving wool to bring out the colour and detail to create a new art from that which had very nearly died out. Today, after three decades of weaving commissions for both Australia and overseas, the Victorian Tapestry Workshop is recognised throughout the world for its quality of craftwork, and for the fact that it is financially secure and weaves better than anybody else anywhere. The people of the Workshop are keeping a craft alive for future generations and for our pleasure today.

Weaver Chris Cochius at the loom creating a wool tapestry commission for the Victorian Tapestry Workshop in South Melbourne.

Unfortunately the proliferation of antique shops and art and craft boutiques throughout the country is exhausting the remaining supply of genuine craft objects without directing proper attention to the need for further 'active and productive' institutions like the Tapestry Workshop. In hundreds of country towns and city suburbs the shelves in reclaimed weatherboard shops are being restocked with inferior and often fake duplicates. Bric-à-brac masquerading as genuine 'bygone' is being freshly manufactured, artificially

Weaving in progress at the Victorian Tapestry Workshop.

General store, Tilba Tilba, New South Wales

aged and sold to eager purchasers. The fact that they *are* eager to buy betrays the nostalgic desire to recapture the better qualities of our past.

The desire has even extended into a kind of amateur archaeology. Some years ago there was a fashion for collecting old bottles. People ranged far and wide in their search for specimens: from city rubbish dumps to the far outback. The famous bottle dump near the Innamincka pub in northern South Australia, which had accumulated its 'treasure' over many years, disappeared within a year or two. And the bottle collectors who used pick and shovel on the local dump began to unearth and extract other 'bygones' in varying degrees of preservation, and without much finesse or care. Most of them were worthless and many were quite unattractive, but nevertheless they had that aura of the past which becomes so important to us in our search for some stable identity. We have lost so much that we are constantly uncertain about the value of its replacements.

One example of our 'vanishing past' is the disappearance of the corner store. The general store, which flourished in our suburbs and country towns is a rarity now. The supermarkets which attract their customers by offering a multitude of 'specials' sell everything at prices the small corner store could not match. The supermarkets have made bulk-buying, mass-marketing and check-outs an artform where the little old general stores had little to offer but individuality and 'just enough'.

Each of the old stores had a flavour of its own, even its own smell. It was in a sense a 'folk theatre' allowing a satisfying interplay between actors: the storekeeper behind his counter and the customers to be greeted and served.

But we can never go backwards. Even if we re-created one old corner store it could only be a museum. We need now to seek other ways of satisfying

A slightly whimsical illustration composed for the endpapers of volumes of the Australian Encyclopaedia *in 1988.*

our need to feel at home in our environment. One way may be by making a deliberate effort to preserve and extend our folklore and fantasy.

We have too little sense of genuine fantasy and celebration in Australia, and what we have is over-commercialised. We lack such 'fantasy events' as Hallowe'en, which the Americans and Germans celebrate each year in October. We have such holidays as Australia Day, as a counterpart to the American Thanksgiving Day, but it has nothing like the same significance. Thanksgiving is a time of family closeness, to celebrate the first harvest brought safely home in a new country. Australia Day is just another casual Aussie holiday at the end of summer, to be marked by a trip to the beach or a cricket match or some other outdoor activity. Many people could not even give a reason for it.

Perhaps the clue lies in that word 'outdoors'. Our climate is usually so genial that we spend as much time outdoors as in. Such celebrations as Christmas and Thanksgiving give the people of the northern hemisphere an opportunity for feasting and gaiety to punctuate the long sombre winters. Our Christmas Days, like so many days in the year, usually have a splendour of sunshine and blue skies which make Christmas trees and roast dinners incongruous.

Another reason for our lack of communal celebration on a national scale may be the absence of genuinely Australian fiestas, or feast days. Most of our fiestas are inherited directly from European ancestors and have little Australian significance. More and more we try and celebrate with comedy, literary and jazz festivals, but they are more self-serving than appealing to a multicultural nation. Some communities celebrate and experiment with festivity, but these usually seem to be restricted to special groups.

'...AND THE BAND PLAYED WALTZING MATILDA'

In recent years the celebration of Anzac Day has been centred far away from Australia at the dawn service at Anzac Cove on Gallipoli in Turkey. C.E.W. Bean, whose words have done more than anybody else's to establish the Anzac tradition, would have been astounded to know that about 300 000 people, most of them young backpackers from Australia and New Zealand, gathered at Anzac Cove for the dawn service on 25 April 2005. Bean returned to the Gallipoli peninsula in 1919, after the war, and as our official war historian recommended that permanent graves be constructed there with pilgrims in mind. The whole Anzac area, he believed, could become the goal of 'pilgrimage from Britain and the Anzac countries'. Until quite recently it was not

'The spirit of mutual reliance which held men together in lonely camps, and then in the formation of unions and on sportsfields, was equally powerful in battle. A man could not let this mates down, and he must go with them into enemy fire just as he would trot with them out onto the football ground. In a sense the war was only another kind of game, with opponents to be beaten so that everyone could go home.'

Michael Page, 1987

easy for aspiring pilgrims to find the transport, time and money to travel all the way across the world from Australia.

Historian Ken Inglis wrote about Anzac Cove in 2005: 'By Anzac Day 1990, the 75th anniversary of the landing, the invaders' descendants had indeed made a place in the holy land. The dawn service addressed by Prime Minister Hawke was attended by perhaps 5000 Australians and New Zealanders and watched at home by millions. Each year since, more people have turned up. Their demeanour and their very presence have created problems such as Bean could never envisage. Some visitors were spending the night before dawn not in reverent meditation but drinking tinnies and belting out songs of their own culture. Should they be offered wholesome entertainment to occupy those hours? John Farnham for example? Custodians of tradition so quickly condemned that idea as a profanation, that it was vetoed by the Anzac Prime Ministers John Howard, who had heard of John Farnham, and Helen Clark, who hadn't.'

No wonder we have problems with festivity and celebration back home where all the Australians are. As Inglis reflects, 'The meanings of Anzac have changed over ninety years, and will go on changing, maybe to accommodate still broader understandings and yearnings by the time of its centenary.'

So perhaps it is time to implement a deliberate conversion in the significance of Anzac Day here in Australia, so that this truly Australian holiday becomes a day for celebration of all that is genuinely Australian in our inheritance and environment. At present, there is a sense of exclusion about the day, in that the greater part of our population cannot share in the memories and experiences of the ex-combatants and those who are related to them. But it has already changed to some degree, in that men and women who did not serve under the Australian flag, even including some of our enemies of long ago, also use it as a meeting day. If, in some way, its meaning could be extended so that it involved the entire population, and became a day to celebrate life as well as remembering death, then it could become Australia's own Thanksgiving Day. It would be a day for celebrating those many things for which we should be grateful: in fact a day of conservation of all that is Australian. The preservation of our nationhood, and those who died for it long ago, would of course be included.

The need for festivity, fantasy and folklore is an important part of our heritage. The three things are very closely associated. Much folklore is fantasy, such as the Aboriginal corroborees. If we seek fantasy, then the Aborigines can provide us with the richest imaginable store. Their myths and legends are in a different form than those of our European origins, but are much more meaningful in the Australian environment. Fortunately, these myths were collected over the years by such men as Bill Harney, C.P. Mountford and Theodore Strehlow. They provide us with an archive of specifically Australian fantasy.

Communities and cultures much older than our own have long since accepted that there is a place for folklore, fairytales and enchantment in the material world of everyday. Australians have been slow to reach this realisation, because so much of their history had to be devoted to the struggle for survival in this 'unchosen land'. Goethe, the German poet who wrote of his mother as a gifted storyteller who wove traditional fairytales into her own stories, said 'From Father I got my bearings, the seriousness of life's pursuits: from Mother the enjoyment of life and the love of spinning fantasies.' Many Australians, new and old, share such feelings, although they are neither bold nor articulate enough to express them. For many, the 'seriousness of life's pursuits' is not balanced by the enjoyment of life in which the imagination plays such an important part. They regard fantasy as escapism, even though we all need to escape.

Even if we cannot 'live happily ever after', at least we may dream of doing so.

We need to encourage people who can open the doors to dreams: encourage magical thinking and the belief in the impossible. If that sounds like a whimsical statement, then it is worth remembering that the mighty multi-national corporations encourage their executives to use 'lateral thinking' and 'brainstorming' for the solution of problems. Such methods specifically partake of fantasy: the search for the 'impossible' which lies behind the 'possible', because it may present a better way of solving a problem. And it should be remembered that every example of modern technology, from television to the space shuttle, once lay in the realm of fantasy and was developed only because some innovator broke the chains of 'the seriousness of life's pursuits'.

We enjoy various types of folklore, festivity and fantasy. Norman Lindsay's *The Magic Pudding*, which is pure fantasy, has been a best loved children's book ever since it was published in 1918. The folklore collections by such anthologists as Bill Wannan and Ron Edwards enjoy large sales, and from time to time there is a renewal of interest in the Henry Lawson stories which are based on our folk traditions.

Australian Rules football and Test cricket are a kind of combination of folklore, folk theatre and festivity. Both perform important functions which are the basis of good theatre. They provide emotional catharsis, and they are enveloped by a mass of complex rituals, seemingly constantly changing, both on the ground and off the ground in the grandstands.

The millions of immigrants and their descendants who have found a place in Australia during the past fifty years have introduced their own inherited festivities, ranging from the Blessing of the Fishing Fleet by the Greek Orthodox church to Morris Dancing in Martin Place, Sydney, by English migrants. Such things are mind-expanding, and since our country is a melting pot of the many different nationalities who have arrived during the past two centuries, there is no reason why Chinese theatre, Latvian dancing and Spanish music should not become part of our festivities, just as the Schützenfeste was introduced to South Australia by Germans who arrived there to settle in 1838. Throughout our cultural history we have accepted things from the Old World and transformed them into something new. If we can preserve the best of the old and change it into something suited to our modern lifestyle, then surely that is one of the best aspects of conservation.

We learn to use hindsight to evaluate our failures and successes, and we patch up our surroundings and our attitudes as our consciences demand. Each generation receives the sum product of all the actions of its forebears. We inherit Australia imprinted with the actions of eight previous generations of white settlement, and the numerous generations before. To those not yet to be born, we bequeath an Australia even more greatly altered.

Our moral obligation is that any changes we make should be in accordance with the laws of nature, and not in defiance of them.

We must maintain Australia's heritage watch, to safeguard the best of all that we have inherited and to raise the alarm whenever our environment is threatened.

BIBLIOGRAPHY

Bates, Daisy, *The Passing of the Aborigines,* Adelaide, 1935.

Bean, C.E.W. Bean, *The Official History of Australia in the War of 1914-18*, Australian War Memorial, Canberra, 1921-1942.

Boyd, Robin, 'The Future of Our Past', in Austrlian National University Proceedings of a Seminar on Historic Preservation in Australia, Canberra, 1967.

Clark, Manning, 'A Discovery of Australia', ABC Boyer Lectures, Sydney, 1976.

Crampton, Patricia, *Silent Night, Holy Night*, Hong Kong, 2005.

Darwin, Charles, *The Origin of Species*, London, 1859.

Egan, Ted, *The Drover's Boy*, Lothian, Melbourne, 1997.

Elliot, Brian, *The Landscape of Australian Poetry*, Melbourne, 1967.

Flinders, Matthew, *Observations on the Coasts of Van Dieman's Land and on Parts of the Coasts of New South Wales*, London, 1802.

Harney, W.E., *Songs of the Songmen*, Adelaide, 1968.

Inglis, Ken, Article, *The Age*, 30 April 2005.

Johnstone, George, *The Australians*, Adelaide, 1966.

Malouf, David, 'A Spirit of Play — Making the Australian Consciousness', ABC Boyer Lectures, Sydney, 1998.

Marsh, George, *Man and Nature*, Cambridge, Mass., 1864.

Miller, Ellen, *The Quiet Land*, Hobart, 1977.

Page, Michael, *The Making of Australians*, J.M. Dent, Melbourne, 1987.

Paterson, A.B., *Clancy of the Overflow*, Sydney, 1889.

Ratcliffe, Francis, *Flying Fox and Drifting Sand*, New York, 1938.

Shapiro, Maurice, *Westernportrait*, Melbourne, 1973.

Thiele, Colin, *Storm Boy*, Rigby, Adleaide, 1974.

Thoreau, Henry, *Walden*, Concord, Mass., 1854.

Wright, Judith, *Australia 1970*, Sydney, 1987

ACKNOWLEDGEMENTS

ILLUSTRATIONS

'Ape Cage, Melbourne Zoo' (p. 47), painting by Robert Ingpen, first published in Ingpen, R., *Marking Time — Australia's Abandoned Buildings*, Rigby, Adelaide, 1979; 'Clancy of the Overflow' illustrations (pp 56-57), by Robert Ingpen, first published in *The Bulletin*, Sydney, 1982; 'A Coral Rock Community' (p. 21), illustration by Robert Ingpen for The Great Barrier Reef Marine Park Authority, 1980; Conservation posters — 'Trees', 'Rivers', 'Forests', 'Wildlife', 'National Parks', 'Coastal Dunes', 'Mangroves', 'Energy' (pp 75-76), by Robert Ingpen, produced for the South Australian Government, 1986; 'The Crags of Capella above Mount Cygnus' (p. 67), painting by Robert Ingpen; first published in Ingpen and Dunkle, *Conservation*, Hill of Content Publishing, 1987; Echidnas illustration (p. 13), from Gould, *The Mammals of Australia*, 1845-63, London; 'El Nino' (p.30), illustration by Robert Ingpen, first published in Ingpen, R., *The Voyage of the Poppykettle*, Rigby, Adelaide, 1981; Endpaper illustration by Robert Ingpen (p. 111), first published in *Australian Encyclopaedia*, Australian Geographic, 1988; First light-weight air raid warning radar installation (p.97), photograph, CSIRO Archive; Illustrations (p. 69), by Robert Ingpen, first published in Ingpen, R., *Westernportrait — An Explanation of the Westernport Bay Environmental Study*, Victorian Ministry for Conservation, 1973; 'Olderfleet Tower' (p. 91), painting by Robert Ingpen, first published in Ingpen and Dunkle, *Conservation*, Hill of Content Publishing, 1987; 'Old River Gum' (p. 23), painting by Robert Ingpen, first published in Ingpen, R., and Thiele, C., *River Murray Mary*, Rigby, Adelaide, 1979; 'Overview of Churchill Island' (p. 71), painting by Robert Ingpen, first published in a brochure of the Victorian Conservation Trust, c. 1980; 'Pioneer Life' postage stamps (pp 50-51), by Robert Ingpen, 1972, courtesy of Postmaster Gallery, Australia Post; 'Shearing Shed' (pp. 98-99), painting by Robert Ingpen, first published in Ingpen, R., *Click Go the Shears*, HarperCollins, 1984; 'Stud Ram Pens' (p. 46), painting by Robert Ingpen, first published in Ingpen, R., *Click Go the Shears*, HarperCollins, 1984; 'Tower Hill' (p. 31),Eugene von Guérard, 1855, courtesy of Warrnambool Art Gallery.

TEXT

Extracts from Egan, Ted, *The Drover's Boy* (pp 61-63), Lothian, Melbourne, 1997; Extracts from Harney, W.E. & Elkin, A.P., *Songs of the Songmen* (pp 54-55 and p. 64), Rigby, 1968; Extract from Lawson, Henry, 'Past Carin" (p. 16), 1899; Extract from Paterson, A.B.,'Clancy of the Overflow' (p. 56); Extract from Paterson, A. B., 'Pioneers' (p.35);.Extract from Wright, Judith, 'Australia 1970' (p. 29), from *A Human Pattern: Selected Poems*, ETT Imprints, Sydney, 1996.